T0259025

Moving Finite Element Method

Fundamentals and Applications in Chemical Engineering

Moving Finite Element Method

Fundamentals and Applications in Chemical Engineering

Maria do Carmo Coimbra
Alirio Egidio Rodrigues
Jaime Duarte Rodrigues
Rui Jorge Mendes Robalo
Rui Manuel Pires Almeida

CRC Press
Taylor & Francis Group
Boca Raton London New York

CRC Press is an imprint of the
Taylor & Francis Group, an **informa** business

MATLAB® is a trademark of The MathWorks, Inc. and is used with permission. The MathWorks does not warrant the accuracy of the text or exercises in this book. This book's use or discussion of MATLAB® software or related products does not constitute endorsement or sponsorship by The MathWorks of a particular pedagogical approach or particular use of the MATLAB® software.

CRC Press
Taylor & Francis Group
6000 Broken Sound Parkway NW, Suite 300
Boca Raton, FL 33487-2742

First issued in paperback 2018

© 2016 by Taylor & Francis Group, LLC
CRC Press is an imprint of Taylor & Francis Group, an Informa business

No claim to original U.S. Government works

ISBN-13: 978-1-4987-2386-2 (hbk)
ISBN-13: 978-1-138-49672-9 (pbk)

This book contains information obtained from authentic and highly regarded sources. Reasonable efforts have been made to publish reliable data and information, but the author and publisher cannot assume responsibility for the validity of all materials or the consequences of their use. The authors and publishers have attempted to trace the copyright holders of all material reproduced in this publication and apologize to copyright holders if permission to publish in this form has not been obtained. If any copyright material has not been acknowledged please write and let us know so we may rectify in any future reprint.

Except as permitted under U.S. Copyright Law, no part of this book may be reprinted, reproduced, transmitted, or utilized in any form by any electronic, mechanical, or other means, now known or hereafter invented, including photocopying, microfilming, and recording, or in any information storage or retrieval system, without written permission from the publishers.

For permission to photocopy or use material electronically from this work, please access www.copyright.com (http://www.copyright.com/) or contact the Copyright Clearance Center, Inc. (CCC), 222 Rosewood Drive, Danvers, MA 01923, 978-750-8400. CCC is a not-for-profit organization that provides licenses and registration for a variety of users. For organizations that have been granted a photocopy license by the CCC, a separate system of payment has been arranged.

Trademark Notice: Product or corporate names may be trademarks or registered trademarks, and are used only for identification and explanation without intent to infringe.

Library of Congress Cataloging-in-Publication Data

Names: Coimbra, Maria do Carmo, author.
Title: Moving finite element method : fundamentals and applications in chemical engineering / Maria do Carmo Coimbra [and 4 others].
Description: Boca Raton : CRC Press, 2016. | Includes bibliographical references and index.
Identifiers: LCCN 2016009190 | ISBN 9781498723862 (acid-free paper)
Subjects: LCSH: Finite element method. | Chemical engineering--Mathematics.
Classification: LCC TP155.2.M36 C65 2016 | DDC 620.001/51825--dc23
LC record available at https://lccn.loc.gov/2016009190

Visit the Taylor & Francis Web site at
http://www.taylorandfrancis.com

and the CRC Press Web site at
http://www.crcpress.com

In memory of
Carlos Alberto Abreu Sereno
3/1/1950 – 29/4/2003
who started the MFEM research at LSRE.

Contents

List of Figures

List of Tables

Symbols

Chapter 1

A	bed cross-section area, m^2
c_i	molar concentration of species i, mol/l
c_{i0}	initial concentration of species i, mol/l
c_{iS}	surface concentration of species i, mol/l
D_e	effective diffusivity, m^2/s
$\tilde{D}_e$	apparent diffusivity, m^2/s
D_h	diffusivity in homogeneous particles, m^2/s
D_i	diffusion coefficient, m^2/s
D_i^0	corrected diffusivity, m^2/s
D^{MS}	Maxwell–Stefan diffusivity, m^2/s
f	friction coefficient, $J.s./(mol.m^2)$
f_i	dimensionless concentration
j_i	diffusion flux, $mol/(s.m^2)$
ℓ	half thickness slab, m
k	fluid thermal conductivity, $J/(m.s.K)$
k	film mass transfer, m/s
$k(T)$	reaction kinetic constant, s^{-1}
k_h	intraparticle LDF mass transfer coefficient, s^{-1}
K	dimensionless constant separation factor
p_i	partial pressure of species i, Pa
q	heat flux x-direction

q_i	adsorbed phase concentration of species i, mol/m^3 adsorbent
$< q_i >$	average adsorbed phase concentration, mol/m^3 adsorbent
q_{i0}	initial adsorbed phase concentration, mol/m^3 adsorbent
q_{iS}	particle surface adsorbed phase concentration, mol/m^3 adsorbent
r	radial particle coordinate, m
R	ideal gas constant, $Jmol^{-1}K^{-1}$
R_p	radius particle, m
T	temperature, K
T	dimensionless throughput parameter of Vermeulen
T_S	catalyst surface temperature, K
u_i	velocity of species i, m/s
u_x	velocity in x direction, m/s
v	bed volume until section z, m^3
v_0	pore velocity, m/s
V	volume of solution passed through a bed section, m^3
x_i	mole fraction of species i
$(-\Delta H)$	heat of reaction, J/mol
ϵ	bed porosity
ϵ_b	interparticle porosity
ϵ_p	particle porosity
Φ	Thiele modulus
η	effectiveness factor of the catalyst

λ_e catalyst thermal effective conductivity

λ_m intraparticle Peclet number

μ_i chemical potential, J/mol

μ viscosity, $Pa.s$

ρ total mass concentration, kg/m^3

ρ_i mass concentration of species i, kg/m^3

τ time constant, s

τ_d diffusion time constant, s

$\tilde{\tau}_d$ apparent diffusion time constant, s

τ_{yx} shearing force per unit area, Pa

ξ_m dimensionless capacity parameter

Chapter 2

$1D$ one-dimensional space domain

$2D$ two-dimensional space domain

$1D + 1d$ space domain with two scales

c_i penalty constants

D_m number of global nodes in Δ_m

$\ell_i^{m,j}(x)$ i-th Lagrange basis function

mP_i i-th micro problem

$\mathcal{M}_m$ spatial mesh associated to u_m

$\mathcal{M}$ spatial mesh in 2D

n number of PDE

n_s number of spatial nodes in $\mathcal{M}$

q_m number of finite elements on $\mathcal{M}_m$

q number of finite elements in $\mathcal{M}$

$p_{m,j}$ number of interpolation points in $\Omega_{m,j}$

r space independent variable in micro domains

R_m residual associated with the m-th PDE

$s_{m,j}^k$ k-th local spatial node in $\Omega_{m,j}$

$S_{m,j}$ internodal spring function in $\Omega_{m,j}$

t time independent variable

tol_1 absolute tolerance for ODE solver

tol_2 relative tolerance for ODE solver

$\mathbf{u}$ exact solution of the system of PDE

u dependent variable for the macro equation

u_p dependent variable for the micro equations

$\mathbf{U}$ MFEM approximation of $\mathbf{u}$

$U_{m,j}$ polynomial approximation to u_m in $\Omega_{m,j}$

$U_{m,j}^i$ value of U_m in the i-th point of the j-th finite element of $\mathcal{M}_m$

$\dot{U}_{m,j}^i$ time derivative of $U_{m,j}^i$

$\tilde{U}_j$ smoothed function in V_δ

V_δ a neighborhood of a spatial node

x space independent variable

$X_{m,j}$ j-th spatial node of the mesh $\mathcal{M}_m$

$\dot{X}_{m,j}$ time derivative of $X_{m,j}$

$Z_{m,j}$ length of the j-th element of $\mathcal{M}_m$

z space independent variable in macro domain

$\varepsilon_{m,j}$ internodal viscosity function in $\Omega_{m,j}$

$\xi_{m,j}^k$ relative positions of interpolation points in the reference interval

Δ_m ordered set of global nodes of $\mathcal{M}_m$

$\Delta_{m,i}$ i-th global node of $\mathcal{M}_m$

$\Phi_{m,i}$ i-th global basis function of $\mathcal{M}_m$

Ω	spatial domain	G_1	dimensionless mass flux at the entry of the top dead volume
$\Omega_{m,j}$	j-th finite element of $\mathcal{M}_m$		
ϑ	link term for macro- and microscales	G	dimensionless mass flux
ϖ	link term in boundary condition for macro- and microscales	G_{ref}	reference mass flux, $kg/(m^2 s)$
		I	dimensionless current externally applied

Chapter 3

		k	reaction constant, s^{-1}
c	dimensionless concentration	$k(T)$	rate constant at temperature T, s^{-1}
c_0	concentration around the particle, mol/m^3	K	dimensionless adsorption constant
c_{feed}	feed concentration, mol/m^3	ℓ	half thickness particle, m
c_A	gas phase concentration of component A, $k.mol/m^3$	L	spatial domain length, reactor length, bed length m
c_p	specific heat for the gas in a constant pressure process, $J/(kgK)$	L_i	length of dead volume i, m
		M_0	initial concentration of the metal, mol/m^3
c_v	specific heat for the gas in a constant volume process, $J/(kgK)$	M_i	molecular mass of species i, g/mol
C_v	valve coefficient	N_d	number of intraparticle mass transfer units
c_i	concentration of component i in the fluid phase, g/l	N_f	number of film mass transfer units by film diffusion
c_{i_0}	feed concentration of component i, g/l	N_w	number of transfer units for the heat transfer at the wall
C_f	heat capacity of the fluid, $J/(kgK)$	p_{feed}	dimensionless feed pressure
C_S	heat capacity of the solid, $J/(kgK)$	p_l	dimensionless initial pressure
d_p	diameter of the particles, m	P	dimensionless total pressure
C_{in}	inlet solute concentration in the gas phase, mol/m^3	P_A	dimensionless partial pressure of species A
$c_i^{m,j}$	penalty constants in the j finite element of mesh $\mathcal{M}_m$	$P_{A_{feed}}$	dimensionless feed partial pressure of species A
$c_6^{m,j}$	minimum admissible length for the j finite element of mesh $\mathcal{M}_m$	Pe	Peclet number
		Pe_{ref}	reference Peclet number
		P_h	high pressure, atm
d	diffusion coefficient	P_l	low pressure, atm
Da	Damköhler number	p_{ref}	reference pressure, atm
D_{ax}	axial dispersion coefficient, m^2/s	P_{BK}	Blake–Kozeny parameter
D_e	effective diffusivity, $m^2 s^{-1}$	q_A	adsorbed phase concentration of component A, $kmol/kg$
E	activation energy, J/mol		

q_i concentration of component i in the solid phase, g/l

q_{in} solute concentration in the adsorbed phase in equilibrium with C_{in}, mol/m^3

Q flow rate, m^3/s

R ideal gas constant, $J/(mol.K)$

R_0 reactor radius, m

S cross-section area of the column, m^2

S_1 cross-section area of the top dead volume, m^2

t dimensionless time variable

tol_1 absolute tolerance for ODE solver

tol_2 relative tolerance for ODE solver

T dimensionless temperature

T_0 initial temperature, K

T_{feed} feed temperature, K

T_w dimensionless wall temperature

u dimensionless dependent variable; electrochemical potential variable

u_i interstitial velocity, m/s

u_s superficial velocity, m/s

U global heat transfer coefficient at the reactor wall

v recovery variable or dimensionless fluid velocity

v_0 intraparticle convective velocity, m/s

V_1 volume of top dead volume, m^3

V_2 volume of bottom dead volume, m^3

v_{ref} reference velocity, m/s

x dimensionless spatial variable

y dimensionless concentration in the adsorbed phase

y_A mole fraction of species A

y_{A_0} initial mole fraction of species A

$y_{A_{feed}}$ feed mole fraction of species A

α stoichiometric coefficient of the impregnant metal

β adiabatic temperature rise

χ ratio between the adsorbed phase concentration in equilibrium with the inlet solute concentration in the gas phase and the initial concentration of the impregnant metal

$(-\Delta H)$ heat of reaction

ϵ bed porosity

ϵ_t total porosity

Φ Thiele modulus

γ Arrhenius number or ratio of specific heat

λ intraparticle Peclet number

μ dimensionless viscosity

μ_B viscosity of species B, Pas

ν flow velocity coefficient

ρ dimensionless fluid density

ρ_{ref} reference fluid density, kg/m^3

ρ_{ap} apparent density of the adsorbent, kg/m^3

ρ_f density of the fluid, kg/m^3

ρ_S density of the solid, kg/m^3

τ space time, s

τ_d diffusion time constant, s

τ_{hl} time constant for thermal wave propagation, s

ξ bed capacity factor

Chapter 4

Bi_h Biot number

C_f heat capacity of the fluid, $J(kg.K)$

C_S heat capacity of the solid, $J(kg.K)$

d	dimensionless diffusion coefficient	c_p	dimensionless intraparticle concentration
G_f	fluid mass flux, $kg/(m^2.s)$	C	bulk concentration, mol/m^3
h_{we}	overall heat transfer coefficient, $J/(m^2.s.K)$	C_0	initial concentration in the bed, mol/m^3
K_{ae}	effective axial thermal conductivity, $J/(m.s.K)$	C_{in}	concentration in the bed, mol/m^3
K_{re}	effective radial thermal conductivity, $J/(m.s.K)$	C_p	intraparticle concentration, mol/m^3
L	bed height, m	D_{ax}	axial dispersion coefficient, m^2/s
Pe_{ha}	thermal axial Peclet number		
Pe_{hr}	thermal radial Peclet number	D_{pe}	effective pore diffusivity, m^2/s
r	radial coordinate, m	k_f	film mass transfer, m/s
R	ideal gas constant, $J.mol^{-1}K^{-1}$	K_L	equilibrium parameter, l/mg
R_c	column radius, m	L	bed height, m
T^*	temperature, K	N_D	number of intraparticle mass transfer units
T	dimensionless temperature		
T_a^*	ambient temperature, $^\circ C$	N_f	film mass transfer units
T_f^*	feed temperature, $^\circ C$	Pe	Peclet number
T_0^*	initial temperature, $^\circ C$	Q	maximum capacity parameter, mg/g
t	dimensionless time variable		
(u,v)	dimensionless dependent variable	r	dimensionless radial coordinate
u_i	interstitial velocity, ms^{-1}		
(x,y)	dimensionless spatial variables	r_i	initial position of spatial nodes in the micro domains
z	axial coordinate, m	R_p	pellets radius, m
β	Prater thermicity factor	u_i	interstitial velocity, m/s
ϵ	bed porosity	z	dimensionless axial coordinate
Φ	Thiele modulus		
γ	Arrhenius number	z_i	initial position of spatial nodes in the macro domain
θ	time variable, s		
ρ_f	density of the fluid, kg/m^3	α	parameter related to the diffusion in the pellets
ρ_S	density of the solid, kg/m^3		
ξ_h	thermal capacity parameter	β	ratio of particle pore volume and interparticle (bed) pore volume

Chapter 5

		ϵ	bed porosity
a_p	particle specific area	ϵ_p	intraparticle porosity
Bi_h	thermal Biot number	ρ_{ap}	apparent density of the adsorbent, g/l
Bi_m	mass Biot number		
c	dimensionless bulk concentration	τ	space time, s
		ξ	adsorbent capacity factor

Chapter 6

ℓ	length of spatial domain
x_c	reduced interface position
s	interface position, m
S	dimensionless interface position
t	dimensionless time variable
t_c	characteristic time, s^{-1}
T	temperature, K
T_f	freezing temperature, K
T_M	melting temperature, K
T_0	temperature at $z = 0$, K
u	dimensionless temperature
x	dimensionless spatial variable
y_i	dimensionless temperature in phase i
z	spatial variable, m
β	dimensionless latent heat
θ	time variable, s

Preface

The vast diversity of problems in science and engineering involves time-dependent partial differential equations (PDE) describing solutions with pronounced spatial activity in some regions of space that move with time. This characteristic of the solution occurs in the simulation of processes involving convection, reaction, diffusion, dispersion, or moving boundaries. In many cases, the complexity of the mathematical models required to describe this kind of problems makes the analytic solution procedures impossible. The numerical simulation is a powerful way to solve those complex models from science and engineering. However, due to the presence of steep moving fronts on the solution and in addition to all the classic requirements, the request for high accuracy in both space and time for all the processes involved is of fundamental importance.

The moving grid methods are designed to concentrate the grid points in regions of high spatial activity of the solution and as the solution is time dependent the grid points are moved in accordance. The moving finite element method (MFEM), introduced by Keith Miller in 1981, is an adaptive grid method in which the grid deforms continuously in time with the solution. Many researchers, among whom the authors themselves are included, recognized a powerful idea in Miller's method and for more than two decades the moving finite elements have been a research area studied in the Laboratory of Separation and Reaction Engineering, LSRE. Thus far, four PhD theses have been written. The research began with the work of Carlos Sereno (1989), a former PhD student of A.E. Rodrigues and J. Villadsen. Sereno formulated and developed the MFEM with polynomial approximations of any degree using Lagrange polynomials to solve a system of explicit convection–diffusion–reaction PDE in a one-dimensional (1D) space domain with fixed endpoints. In this formulation, each variable has its own moving grid and at each instant the MFEM computes the numerical solution of the PDE system and also the spatial grids where each component is represented. Maria C. Coimbra (2000) extended the work of Sereno and developed a formulation to solve implicit systems of convection–diffusion–reaction PDE. Moreover, she developed an algorithm for evolutionary problems in two-dimensional (2D) space domains with fixed boundaries. Another LSRE researcher, Jaime Rodrigues (2010), proposed a strategy designed for solving time-dependent differential equations involving one or more changes of phase. These problems are often called moving boundary problems or Stefan problems. He also addressed evolutionary problems in spatial domains of higher dimension, taking an interest in problems involving two scales (1D+1d). Recently, Rui Robalo (2014) has focused his research on the moving finite element method to solve Stefan problems with one or more interfaces in two-dimensional space domains. From the numerical algorithms based on MFEM, computer codes were developed either in FORTRAN or MATLAB® that allow the numerical simulation of such problems.

The core of the book focuses on process simulation in chemical engineering with a numerical algorithm based on the moving finite element method. It offers new tools and approaches for modeling and simulation problems with moving fronts. These problems are described by time-dependent convection–reaction–diffusion partial differential equations in a fixed spatial domain or in a spatial domain with one or more moving boundaries. The book provides a comprehensive account of the development of the moving finite element method, describing and analyzing the theoretical and practical aspects of the moving finite element method for models in 1D, 1D+1d, and 2D space domains. Relevant numerical applications of the moving finite element method from real-world process simulation are presented. The features of the computational packages based on the moving finite element method are shown and discussed.

The outline of the book is as follows.

Chapter 1 offers some reflections on modeling and simulation in chemical engineering (ChE). It starts with an overview of the evolution of ChE, from the unit operation idea to the movement of chemical engineering science and today's vision focused on process and product engineering, and molecular and materials engineering. The philosophy of modeling is addressed in its aspects of model definition, simplifications, and strategy of modeling. The "art"of modeling is discussed to emphasize the importance of scaling, dimensionless groups, choice of variables, and connection of models with real life. Examples of $10 models (equilibrium model of fixed-bed adsorbers), $100 models (convection, diffusion, and reaction in large pore catalysts), and $1000 models (CFD modeling of fluid flow in reactors) are presented from different areas of ChE. Finally two examples of process simulation are discussed from perfusion chromatography and CO_2 capture by adsorption.

Chapter 2 presents a description of the moving finite element method. We are particularly interested in applying the moving finite element method to solve time-dependent partial differential equations or systems in one- or two-dimensional space domains that have solutions exhibiting sharp moving fronts. These sharp moving fronts can occur in problems with fixed boundaries or in problems with moving boundaries, called Stefan problems. First, the governing mathematical equations are introduced to find the solution of different kinds of time-dependent partial differential equations with sharp moving fronts. Second, the fundamental properties of the moving finite element method in one-dimensional space domains with fixed endpoints are presented in detail. Then, the formulation of the moving finite element method for higher dimensions is considered. Two different models are treated. One designed for real-life processes with two different space scales, a macroscale, and a microscale connected by one of the boundaries of the microscale. The other is created to deal with evolutionary problems in the two-dimensional space domain. Finally, the formulation of the moving finite element method that applies to moving boundary problems in 1D or 2D spatial domains is proposed. The algorithm can deal with moving boundary problems with one or more moving interfaces. For these multiphase problems, the moving finite element algorithm determines the positions of all moving interfaces along time, together with the solution. Also, nonlinear boundary conditions at these moving interfaces are admitted.

The second part of the book, consisting of Chapters 3 to 6, concerns numerical simulation and covers a broad range of application algorithms to engineering problems, namely, in separation and reaction processes.

Chapter 3 begins with a description and implementation of the algorithm, based on the moving finite element method, to solve one-dimensional space domain problems. Applications addressed include some problems presented in the literature as test problems and others from chemical engineering such as the model from the nonisothermal tubular catalytic reactor or the separation of optical isomers by chromatography, or the simulation of the pressurization phase of pressure swing adsorption processes.

Chapter 4 deals with numerical simulations of evolutionary problems in two-dimensional space domains with fixed boundaries. The computational package, based on the moving finite element method, allows numerical simulation of scalar models or systems of time-dependent PDE models. Numerical simulations of a reaction–diffusion equation, Burgers' equation, and Burgers' system and a problem from heat transfer in a fixed bed are presented after a brief introduction and description of the numerical code.

Chapter 5 is concerned with the numerical simulation of problems with the presence of two different spatial scales, a macroscale and a microscale linked by one of the boundaries of the microscale. It is the case of the simulation of a fixed-bed adsorption process in which the macroscale can be associated with the fixed bed and the microscale to the adsorbent particles.

Aiming to expand the range of applications, Chapter 6 shows the numerical simulation of the moving finite element method that applies to moving boundaries or Stefan problems in one- or two-dimensional space domains. The strategy implemented is described, and the features of the computer package are analyzed. A range of issues have been considered including solidification and melting processes, ice–water–ice Stefan problem and isothermal and nonisothermal shrinking core model with nonlinear kinetics. The shrinking core model for the discharge of a metal hydride electrodes problem, causticizing reaction simulation for the Kraft paper process of pulp production, a model for swelling of wood grease in a wool scour, and the Frank spheres problem are also simulated.

The last chapter sums up our findings and outlines potential areas of future research and application of the moving finite element method.

MATLAB®, is a registered trademark of The MathWorks, Inc. For product information please contact:

The MathWorks, Inc.
3 Apple Hill Drive
Natick, MA, 01760-2098 USA
Tel: 508-647-7000
Fax: 508-647-7001
E-mail: info@mathworks.com
Web: www.mathworks.com

Contributors

Maria do Carmo Coimbra
Associate Laboratory LSRE — Laboratory of Separation and Reaction
   Engineering
Department of Civil Engineering
University of Porto
Porto, Portugal

Alírio E. Rodrigues
Associate Laboratory LSRE — Laboratory of Separation and Reaction
   Engineering
Department of Chemical Engineering
University of Porto
Porto, Portugal

Rui J. Robalo
Center for Mathematics and Applications
University of Beira Interior
Covilhã, Portugal

Jaime D. Rodrigues
Department of Mechanical Engineering
University of Porto
Porto, Portugal

Rui Almeida
Center for Mathematics and Applications
University of Beira Interior
Covilhã, Portugal

1

Modeling and Simulation in Chemical Engineering

1.1 The Evolution of Chemical Engineering (ChE)

In this section we will briefly describe the evolution of chemical engineering (ChE) following the advice of Dwig Prater given to Jim Wei [1] on how to present a communication: "Tell them what you will tell them, then you tell them, and finally you tell them what you have just told them." We start with the paradigms of ChE (unit operations and chemical engineering science movement) and finish with our view of today's ChE.

1.1.1 The first paradigm of chemical engineering: Unit operation

In the sixties when studying ChE one still had a course describing specific industries such as sulfuric acid and ammonia. This approach can be viewed as the pre-paradigm of ChE; the evolution of ChE is well described in the so-called Amundson Report [2].

The first paradigm — unit operations — starts really with George E. Davis (1850–1906). In a series of lectures on operation of chemical processes at Manchester Technical School in 1887, he used a "unit operations" approach with focus on pilot scale tests and safety rules. The term "chemical engineering" to designate the profession was used for the first time.

Lewis Mills Norton (1855–1893) taught at MIT in 1888 the first 4–year course in chemical engineering — "Course X." The first chemical engineer graduated in 1891, William Page Bryant.

William H. Walker (1869–1934), Warren K. Lewis (1882–1975), and Arthur D. Little (1863–1935) are the pioneers who defined chemical engineering as a profession with proper approach and *training* methods. Arthur D. Little was the first to use the term "unit operation" in a report to the president of MIT (1915). Walker restructured the curriculum of ChE at MIT and in 1908 the Research Laboratory of Applied Chemistry was created. Warren K. Lewis became a staff member of MIT in 1908; his ability to theorize engineering problems and his strong character contributed to strengthen the program.

The teaching of unit operations became quantitative. This period is condensed in the book: *The Principles of Chemical Engineering* (1923) by Walker, Lewis, and William H. McAdams [3].

1.1.2   The second paradigm of ChE: Chemical engineering science movement

The chemical engineering science movement is illustrated by the book of Bird, Stewart, and Lightfoot, *Transport Phenomena* [4] (see Figure 1.1), which treats transport processes of momentum, heat and mass, which are after all a major portion of the ChE activity. The transport phenomena approach was initiated by Kramers (Delft University of Technology) where Bird spent a semester and knew the lecture notes *Physische Transportverschijnselen.*

FIGURE 1.1
The chemical engineering science movement illustrated by the book of Bird, Stewart, and Lightfoot, *Transport Phenomena.*

The basic laws of momentum, heat conduction, and mass transfer by diffusion are Newton's law, Fourier's law, and Fick's law, respectively:
   Momentum transfer — Newton's law:

$$\tau_{yx} = -\mu \frac{du_x}{dy} \tag{1.1}$$

where μ is the viscosity, u_x is the velocity in x direction, τ_{yx} is the shearing force per unit area.
   Heat transfer by conduction — Fourier's law:

$$q = -k \frac{dT}{dx} \tag{1.2}$$

where q is the heat flux in the x-direction, k is the thermal conductivity, and T is the temperature.
   Mass transfer by diffusion — Fick's law:

$$j_i = -D_i \frac{d\rho_i}{dx} \text{ at constant } \rho \tag{1.3}$$

where j_i is the diffusion flux, D_i is the diffusion coefficient, ρ_i is the mass concentration of species i, and ρ is the total mass concentration.

The above laws assume infinite velocity of propagation of the signal. This problem is eliminated following the proposal of James Clerk Maxwell [5] for momentum transfer:

$$\tau_{yx} + \tau \frac{\partial \tau_{yx}}{\partial t} = -\mu \frac{du_x}{dy} \qquad (1.4)$$

where τ is the time constant and τ_{yx} is the shear stress in a fluid or solid body;

$$j_i + \tau \frac{\partial j_i}{\partial t} = -D_i \frac{d\rho_i}{dx} \qquad (1.5)$$

for mass transfer [6] and

$$q + \tau \frac{\partial q}{\partial t} = -k \frac{dT}{dx} \qquad (1.6)$$

for heat transfer (VC equation of Vernott and Cattaneo) [7]. In homogeneous substances, the relaxation time is $10^{-8} - 10^{-14}$ s and Fourier's law works for normal heating processes. But in biological systems τ is of the order of $10-30$s and VC equation applies.

1.1.3   Chemical engineering today

Solke Bruin in his inaugural lecture at the TU Eindhoven, "Product-driven process engineering"[8], recognizes 4 major classes of processes: Separation processes, structuring processes, transformation processes, and stabilization processes. He makes a point in one statement: Chemical engineering is not anymore just "To make" but more importantly "To service" and "To care."

Product engineering (or product design) entered chemical engineering curricula as a response to the changing nature of jobs for chemical engineers, shifting from 75% in commodities in 1975 to 50% in products in 2010. The teaching of product engineering has been supported by the books of Cussler and Moggridge [9], Ulrich and Eppinger [10], Wesselingh et al. [11], and Wei [12].

Our view of today's ChE [13] is represented by the diamond with corners of process and product engineering, materials and molecular engineering or $ChE = M2P2E$ in short (Figure 1.2).

Process engineering focuses on the synthesis, design, operation, control, and optimization of plants, while product engineering aims at meeting consumer needs by making products with desired function. Materials engineering is present in classical industries (adsorbents, catalysts) but also in many new areas of electronics, photovoltaics, etc. The quest for understanding processes and products at the molecular level is the driving force for the development of molecular engineering.

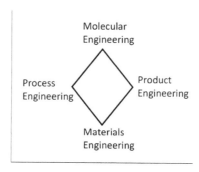

FIGURE 1.2
The chemical engineering diamond [13].

1.2   Philosophy of Process Modeling

One of the authors when studying in Nancy in the early seventies remembers Professor Pierre Le Goff saying "Le Génie Chimique c'est pas de la plomberie." He used to say that a chemical engineer when solving a problem writes [14]:

i) conservation equations
(mass, energy, momentum, electric charge)

ii) equilibrium law at the interface (s)

iii) constitutive laws (for example, ideal gas law)

iv) kinetic laws of transport (heat/mass) and reaction

v) initial and boundary conditions

vi) optimization criteria

This methodology has been useful to analyze problems at multiple scales involved in ChE: pore scale (catalyst, adsorbent) : 1nm − 1000nm; particle scale : 10μm − 1cm; reactor/separator scale: 1m − 10m.

1.2.1   What is a model?

According to Aris [15] a "mathematical model" or simply model "is a complete and consistent set of mathematical equations which are supposed to correspond to some entity — its prototype — which can be a physical, biological, social . . . entity although here we deal with physicochemical systems."

A process model is a relation between "outputs" and "inputs" (feed conditions, design parameters, process adjustable parameters; Shinnar [16]) in view of: i) scale-up from lab to industrial scale; ii) prediction of process dynamics, and iii) optimization of operating conditions.

Models can be classified in many ways: deterministic vs. probabilistic; linear vs. nonlinear; steady state vs. nonsteady state or lumped parameter (perfectly mixed reactor) vs. distributed parameter (fixed-bed reactor) [17].

1.2.2 Models: Simplification of reality, to "better" know the reality

The detail of mathematical description can be guided by objectives which can seem contradictory.

One step in modeling is the *simplification of reality or idealization*. In an excellent paper Levenspiel [18], a pioneer of chemical reaction engineering, mentions Denbigh [19]: "In science it is always necessary to abstract from the complexity of the real world, and in its place to substitute a more or less idealized situation that is more amenable to analysis." This idealization leads to the creation of new models, simplified, which are a "digital impression" of our profession. Examples are: i) boundary layer theory; ii) model of film heat transfer; h; iii) model of film mass transfer; k; iv) theory of residence time distribution (RTD) and tracer technology.

A second step is *detailed modeling to "better" know the reality*. This can be illustrated by the Maxwell–Stefan model for multicomponent diffusion [20]. The driving force is the gradient of chemical potential, μ_i [21]. The diffusive flux due to this gradient is balanced by friction forces

$$f u_i = -\frac{d\mu_i}{dx} \tag{1.7}$$

where f is the friction coefficient and u_i is the velocity of species i. The flux is

$$N_i = u_i c_i - \frac{RT}{f}\frac{\partial \ln p_i}{\partial \ln c_i}\frac{dc_i}{dx} \tag{1.8}$$

where c_i is the molar concentration of species i, the "corrected diffusivity" is $D_i^0 = \frac{RT}{f}$ and the thermodynamic factor is $\frac{\partial \ln p_i}{\partial \ln c_i}$.

For binary systems if the change in partial pressure of species 1 is $-dp_1$ over distance x, the force acting in 1 per volume is $-\frac{dp_1}{dx}$. If the concentration of 1 is c_1 the force per mole of species 1 is $-\frac{1}{c_1}\frac{dp_1}{dx}$. For an ideal gas,

$$-\frac{RT}{p_1}\frac{dp_1}{dx} = -RT\frac{d \ln p_1}{dx} = -\frac{d\mu_1}{dx} \tag{1.9}$$

This force is balanced by the friction between species 1 and 2, proportional to the difference of velocities and to the concentration of component 2, expressed by its mole fraction x_2. The balance of forces acting in species 1 is

$$-\frac{d\mu_1}{dz} = \frac{RT}{D^{\mathrm{MS}}}x_1 x_2(u_1 - u_2) \tag{1.10}$$

or

$$-\frac{1}{P}\frac{dp_1}{dz} = \frac{1}{D^{\mathrm{MS}}}x_1 x_2(u_1 - u_2) \tag{1.11}$$

For n components,

$$d_i = \nabla x_i = \sum_{j=1}^{n}\frac{x_i N_j - x_j N_i}{c_t D_{ij}^{\mathrm{MS}}} \tag{1.12}$$

1.2.3   Strategy of modeling

A philosophy of modeling can be based in 4 points [22]: i) Start with simple models; obtain from such models information which remains valid for more complex models (\$10 approach of Levenspiel: "Always start by trying the simplest model and then only add complexity to the extent needed"); ii) The validity of a model is not just a result of a "good fit;" more important is the capability to predict the system behavior under operating conditions different from those used to get model parameters; iii) Good results can only be obtained if the model "well" represents the system; iv) Use models to obtain useful design parameters and their dependence on operating conditions; use independent experiments if possible to get model parameters.

In short: model development is a task to be carefully done to avoid waste of energy in the next simulation step. "Keep things as simple as possible, but not simpler" (Einstein).

1.3   The "Art" of Modeling

The "art" of modeling uses some techniques such as non-dimensionalization and scaling, averaging, appropriate choice of independent variables [23, 24].

1.3.1   Scaling, dimensionless groups, averaging, choice of variables

Chemical engineers are used to making normalization of variables. As a consequence of that mathematical operation dimensionless groups appear with a physical meaning. A simple example is the problem of diffusion/reaction in an isothermal porous catalyst with slab geometry. The mass balance in steady state for irreversible reaction of order n is:

$$D_e\frac{d^2 c_i}{dz} - kc_i^n = 0 \tag{1.13}$$

with boundary conditions (symmetry condition in the center and surface condition)

$$z = 0, \ \frac{dc_i}{dz} = 0$$
$$z = \ell, \ c_i = c_{iS} \tag{1.14}$$

The normalization of the space variable, z, by the half thickness of the slab ℓ and concentration variable by the surface concentration leads to

$$\frac{d^2 f_i}{dx^2} - \ell^2 \frac{k c_{iS}^{n-1}}{D_e} f_i^n = 0 \tag{1.15}$$

which shows the dimensionless group governing the reaction–diffusion problem:

$$\ell^2 \frac{k c_{iS}^{n-1}}{D_e} = \Phi^2 \tag{1.16}$$

where Φ is the Thiele modulus.

The physical meaning of the dimensionless group is: Φ^2 is simply reaction rate/diffusion rate or diffusion time constant/ reaction time constant. Two extreme cases arise: i) reaction rate $<<$ diffusion rate: the concentration profile inside the catalyst is almost equal to the surface concentration; the catalyst works on "chemical regime"; ii) reaction rate $>>$ diffusion rate: steep concentration profile near the catalyst surface; the catalyst works in "diffusional regime."

Another technique often used by chemical engineers is averaging. In many situations, one does not need the detailed concentration profile inside an adsorbent or catalyst particle but only the average concentration.

Let us illustrate this technique with the LDF model (linear driving force) of Glueckauf [25] for a spherical "homogeneous" adsorbent particle (Figure 1.3). The mass balance in a volume element of the particle located between spheres of radius R and $R + dR$ is:

$$\frac{\partial q_i}{\partial t} = D_h \frac{1}{R^2} \frac{\partial}{\partial R} \left(R^2 \frac{\partial q_i}{\partial R} \right) \tag{1.17}$$

where q_i is the adsorbed phase concentration of species i and D_h is the diffusivity in "homogeneous" particles. The boundary conditions are: at the center of the particle a zero flux condition or symmetry condition and, at the particle surface the adsorption equilibrium isotherm $q_{iS} = f(c_i)$ links the adsorbed phase concentration q_{iS} and the fluid concentration:

$$R = 0, \ \frac{\partial q_i}{\partial R} = 0$$
$$R = R_p, \ q_{iS} = f(c_i) \tag{1.18}$$

The averaging operation consists of multiplying both members by $R^2 dR$, and

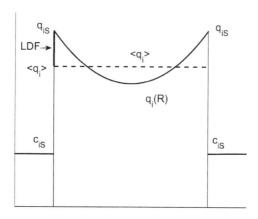

FIGURE 1.3
Linear driving force (LDF) model.

integrating over the particle volume (between 0 and R_p) and introducing average concentration $< q_i >$; the result is for a parabolic profile of adsorbed phase concentration:

$$\frac{\partial < q_i >}{\partial t} = \frac{15 D_h}{R_p^2}\left(q_{iS} - < q_i >\right) = k_h\left(q_{iS} - < q_i >\right) \qquad (1.19)$$

The right choice of variables can help in process modeling. Let us consider the simple model of fixed-bed adsorption: the so-called equilibrium model which assumes isothermal operation, plug fluid flow, diluted systems (*trace system*), negligible pressure drop.

Model equations: species mass balance in a bed volume element and the equilibrium law at the fluid/solid interface:

$$u_0\frac{\partial c_i}{dz} + \epsilon\frac{\partial c_i}{\partial t} + (1 - \epsilon)\frac{\partial q_i^*}{\partial t} = 0$$
$$q_i^* = f(c_i) \qquad (1.20)$$

Normalizing the dependent variables, $\tilde{c}_i = \dfrac{c_i}{c_{i0}}$ and $\tilde{q}_i = \dfrac{q_i^*}{q_{i0}^*}$ we get for an adsorption equilibrium isotherm of "constant separation factor" type,

$$u_i\frac{\partial \tilde{c}_i}{\partial z} + \frac{\partial \tilde{c}_i}{\partial t} + \frac{1 - \epsilon}{\epsilon}\frac{q_{i0}^*}{c_{i0}}\frac{\partial \tilde{q}_i}{\partial t} = 0$$
$$\tilde{q}_i = \frac{K\tilde{c}_i}{1 + (K - 1)\tilde{c}_i} \qquad (1.21)$$

A first dimensionless parameter appears: the "capacity parameter" of the adsorption column

$$\xi_m = \frac{1-\epsilon}{\epsilon}\frac{q_{i0}^*}{c_{i0}} \tag{1.22}$$

Furthermore we can combine the independent variables z and t in only one variable T (throughput parameter of Vermeulen [26]) defined as the ratio of moles of solute passed through the bed section located at $v = Az$ and the number of moles retained in the adsorbent contained in the volume v, is:

$$T = \frac{c_{i0}(V - \epsilon v)}{(1-\epsilon)vq_{i0}^*} = \frac{1}{\xi_m}\left(\frac{u_i t}{z} - 1\right) \tag{1.23}$$

Now the mass balance becomes simply:

$$\frac{d\tilde{q}_i}{d\tilde{c}_i} = T \tag{1.24}$$

Moreover, introducing the derivative of the adsorption equilibrium isotherm we obtain the breakthrough curve for unfavorable isotherms as shown in Figure 1.4,

$$\tilde{c}_i = \frac{1 - \sqrt{\frac{K}{T}}}{1 - K}, \quad K \leq T \leq 1/K \tag{1.25}$$

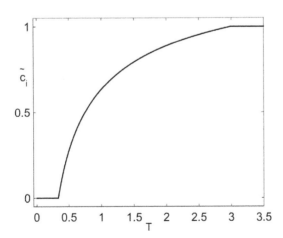

FIGURE 1.4
Breakthrough curve for unfavorable adsorption isotherms from equilibrium model of fixed-bed columns for $K = \frac{1}{3}$.

1.3.2 From model results to real life

Let us go back to the reaction–diffusion problem for first-order reaction in isothermal slab catalyst particle in Section 1.3.1. The concentration profile is

$$f_i = \frac{\cosh(\Phi x)}{\cosh(\Phi)} \tag{1.26}$$

The effectiveness factor of the catalyst (ratio between the observed rate and the reaction rate at the surface condition) is:

$$\eta = \frac{\tanh(\Phi)}{\Phi} \tag{1.27}$$

Knowledge of the effectiveness factor is important to calculate the amount of catalyst needed to get a given reactant conversion in the reactor. The problem is that to calculate the Thiele modulus the kinetic constant k must be known, which is not the case many times in real life. However, theoretical results can be transformed in practical tools. Weisz and Prater [27] changed the $\eta = f(\Phi)$ plot in a more useful η versus $\eta\Phi^2$ plot since the last group does not require the knowledge of k but only observable quantities,

$$\eta\Phi^2 = \frac{r_{obs}\ell^2}{c_{iS}D_e} \tag{1.28}$$

Another aspect is the obtention of useful relations between dependent variables that one can get from proper modeling. One example is diffusion/reaction/conduction in a nonisothermal slab catalyst and first-order irreversible reaction. Model equations include mass/energy balances:

$$D_e \frac{d^2 c_i}{dz^2} - k(T)c_i = 0$$

$$\lambda_e \frac{d^2 T}{dz^2} + (-\Delta H)k(T)c_i = 0 \tag{1.29}$$

$$z = 0, \ \frac{dc_i}{dz} = \frac{dT}{dz} = 0$$

$$z = \ell, \ c_i = c_{iS}; \ T = T_S$$

With some manipulation (multiplying the first equation by the heat of reaction $(-\Delta H)$, adding the second and integrating twice the resulting equation) we get the Damköhler equation [28] relating concentration and temperature inside the catalyst,

$$T - T_S = \frac{D_e(-\Delta H)}{\lambda_e}\left(c_{iS} - c_i\right) \tag{1.30}$$

1.4 Model Classification of Levenspiel: \$10, \$100, and \$1000 Models

1.4.1 \$10 models: Equilibrium models of adsorption columns

The model assumes isothermal operation, plug fluid flow, infinitely fast mass transfer between fluid and solid phases (instantaneous equilibrium at the interface), and trace system. Model equations are Equations 1.20, i.e., the mass balance and the equilibrium law at the interface fluid/solid. Using the cyclic relation between partial derivatives we get the De Vault equation [29],

$$u_{c_i} = \left.\frac{\partial z}{\partial t}\right|_{c_i} = \frac{u_i}{1 + \frac{1-\epsilon}{\epsilon} f'(c_i)} \tag{1.31}$$

This is the most important result to retain in understanding adsorptive and chromatographic processes as a phenomenon of propagation of concentration waves. The simplest model shows that the nature of the equilibrium isotherm is the main factor influencing the shape of the breakthrough curve. The physical concepts to be retained are: dispersive waves are formed when isotherms are unfavorable; each concentration propagates with a velocity given by De Vault equation. Compressive waves are formed for favorable isotherms and, the physical limit is a shock which propagates with a velocity

$$u_{sh} = \frac{u_i}{1 + \frac{1-\epsilon}{\epsilon} \frac{\Delta q_i}{\Delta c_i}} \tag{1.32}$$

where the slope of the chord linking the feed state and the bed initial state appear instead of the local slope of the equilibrium isotherm (Figure 1.5).

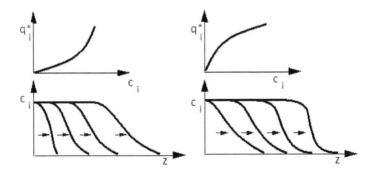

FIGURE 1.5
Unfavorable isotherms and dispersive fronts (left), favorable isotherm and compressive front (right).

1.4.2 $100 models: Diffusion, convection, and reaction in permeable catalysts

This problem was analyzed in detail by Nir and Pismen [30] for first-order irreversible reaction in isothermal catalysts. However, the problem was first tackled by Wheeler in 1954 [31] who concluded that intraparticle convection would be important only for gas phase systems at high pressure in catalysts with very large pores. For the reaction $A \rightarrow B$ in slab isothermal catalysts the mass conservation equation is:

$$\frac{d^2 f_A}{dx^2} - 2\lambda_m \frac{df_A}{dx} - 4\phi_S^2 f_A = 0 \tag{1.33}$$

with boundary conditions, $f_A = 1$ at $x = 0$ and $x = 1$. Model parameters are now the Thiele modulus $\phi_S = \ell\sqrt{\frac{k}{D_e}}$, (ϕ_S^2 ratio between time constants for pore diffusion and reaction); and the intraparticle Peclet number $\lambda_m = \frac{v_0 \ell}{D_e}$ (ratio between time constants for pore diffusion and convection). The concentration profile inside the catalyst is:

$$f_A = \frac{\sinh(\alpha_2)e^{\alpha_1(2x-1)} - \sinh(\alpha_1)e^{\alpha_2(2x-1)}}{\sinh\left(\alpha_2 - \alpha_1\right)} \tag{1.34}$$

where

$$\alpha_{1,2} = \frac{1}{2}\left(\lambda_m \pm \sqrt{\lambda_m^2 + 4\Phi_S^2}\right) \tag{1.35}$$

Interparticle concentration profiles are shown in Figure 1.6.
The effectiveness factor is

$$\eta_{dc} = \frac{\frac{1}{\alpha_1} - \frac{1}{\alpha_2}}{\coth\alpha_1 - \coth\alpha_2} \tag{1.36}$$

When convection is not important, i.e., $\lambda_m = 0$, then $\eta_d = \frac{\tanh\Phi_S}{\Phi_S}$. The effect of convection can be seen in the intermediate region of Thiele modulus (similar reaction and diffusion rates) where the effectiveness of the catalyst is improved by convection. The pore convection will apparently increase diffusivity and move the catalyst working regime from diffusional to "chemical" controlled. The message is: intuition is not enough!

1.4.3 $1000 models: CFD simulation of fluid flow in chemical reactors/separators

Danckwerts (1953) [32] approached the study of fluid flow in reactors in a simple way: "introduce a pulse of tracer into the fluid entering the reactor and see when it leaves"(see Figure 1.7). The normalized outlet concentration versus time is the residence time distribution (RTD). The study of RTD of

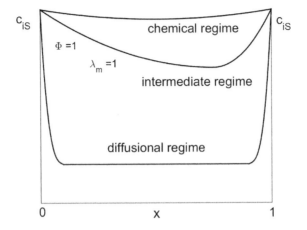

FIGURE 1.6
Asymmetric concentration profiles can exist in convection, diffusion, and reaction in catalyst particles.

flowing fluids can be put under the umbrella of tracer technology, which is relevant not only for chemical engineers, but also for researchers in the medical field, environment, etc., to diagnose the reactor behavior, drug distribution in the body, etc. Danckwerts built a theory based on the characterization of fluid elements of a population inside the reactor (age and life expectation) and leaving the reactor (residence time). Then he introduced the "distribution": relative to each character; the residence time distribution $E(t)$ is then defined as $E(t)dt$ being the fraction of fluid elements leaving the reactor with residence time between t and $t + dt$. The next question is how to experimentally have access to $E(t)$. This brings the tracer technology to the center of the arena. The normalized response to an impulse of tracer $C(t)$ is directly related with the RTD, i.e., $C(t) = \tau E(t)$; or the normalized response to a step input of tracer $F(t)$ curve of Danckwerts is $E(t) = \frac{dF(t)}{dt}$.

Levenspiel [18] summarizes the progress on the study of fluid flow: "In the 19th century there were two approaches to study fluid flow: hydrodynamics (dealt with ideal frictionless fluid; highly mathematical stuff) and hydraulics developed by civil engineers "who amassed mountains of tables of pressure drop and head loss of fluids in open and closed channels of all sort .... " At the beginning of the 20th century Prandtl said "Hydrodynamics has little significance for the engineer because of the great mathematical knowledge required for an understanding of it and the negligible possibility of applying its results. Therefore, engineers put their trust in the mass of empirical data collectively known as the "science of hydraulics." Prandtl was the genius who

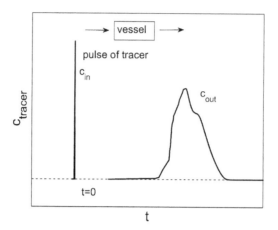

FIGURE 1.7
Tracer experiment and residence time distribution.

patched together these different disciplines with his simple boundary layer theory. The result is modern fluid mechanics.

On the other hand, numerical methods for the solution of PDEs exist and the combination of two solid disciplines appears with a new name: "Computational Fluid Dynamics." Many years ago one of us published in ISCRE8 "Residence time distribution in laminar flow through reservoirs from momentum and mass transport equations" [33]. It is a problem of 2-D flow in a reservoir of length L and height H where a stationary laminar flow exists between inlet and outlet (Figure 1.8). The formulation is made in terms of continuity equation, momentum equations, and species transport equation; the flow field is calculated and the RTD is obtained by solving the mass conservation equation:

$$\frac{\partial(u_x)}{\partial x} + \frac{\partial(u_y)}{\partial y} = 0$$

$$\rho\frac{\partial(u_x)}{\partial t} + \rho\left(\frac{\partial u_x^2}{\partial x} + \frac{\partial(u_x u_y)}{\partial y}\right) = -\frac{\partial P}{\partial x} + \mu\left(\frac{\partial^2 u_x}{\partial x^2} + \frac{\partial^2 u_x}{\partial y^2}\right)$$

$$\rho\frac{\partial(u_y)}{\partial t} + \rho\left(\frac{\partial(u_x u_y)}{\partial x} + \frac{\partial u_y^2}{\partial y}\right) = -\frac{\partial P}{\partial y} + \mu\left(\frac{\partial^2 u_y}{\partial x^2} + \frac{\partial^2 u_y}{\partial y^2}\right) \qquad (1.37)$$

$$\frac{\partial c}{\partial t} + \frac{\partial(u_x c)}{\partial x} + \frac{\partial(u_y c)}{\partial y} = D\left(\frac{\partial^2 c}{\partial x^2} + \frac{\partial^2 c}{\partial y^2}\right)$$

This problem was solved with modern CFD tools (Fluent) more recently [34] and shown in Figure 1.8 for various times normalized by the space time τ.

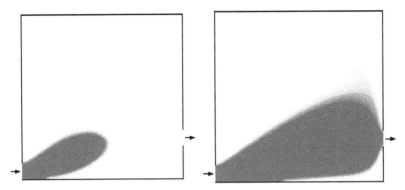

FIGURE 1.8
2-D flow in a reservoir and RTD, $\theta = t/\tau = 0.1$ (left) and $\theta = 1$ (right) for $Re = 10$ and $L/H = 1$ (adapted from [34].)

1.5 Process Simulation

1.5.1 Perfusion chromatography

The factors influencing the behavior of a fixed bed chromatographic column can be classified in two categories: equilibrium and kinetic factors (hydrodynamics, heat/mass transfer). When chromatographic packings contain large pores (> 1000 A) for transport and smaller pores to provide adsorption capacity we enter the domain of perfusion chromatography [35] (Figure 1.9). Mass transfer inside particles occurs then by diffusion and convection.

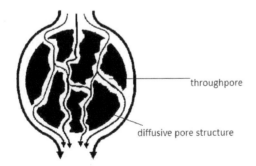

throughpore

diffusive pore structure

FIGURE 1.9
Large-pore material for perfusion chromatography.

Assuming intraparticle mass transport not only by diffusion, D_e, but also by convection (pore velocity v_0) and the equivalence with the conventional model where both mechanisms are lumped in an apparent diffusivity $\tilde{D}_e$ leads to [36]

$$\tilde{D}_e = D_e \frac{1}{f(\lambda)} \tag{1.38}$$

where the intraparticle Peclet number is $\lambda = \dfrac{v_0 \ell}{D_e}$ for a packing with characteristic dimension ℓ. The apparent diffusivity is augmented by convection, and the enhancement factor is

$$\frac{1}{f(\lambda)} \tag{1.39}$$

This result explains the functioning of perfusion chromatography which appeared in 1990.

For a nonadsorbable tracer the "lumped" diffusion-convection model for transient state is:

$$\tilde{D}_e \frac{\partial^2 c}{\partial x^2} = \epsilon_p \frac{\partial c}{\partial t} \tag{1.40}$$

The particle transfer function is:

$$\tilde{g}_p(s) = \frac{<\bar{c}>}{\bar{c}_s} = \frac{\tanh \sqrt{\tilde{\tau}_d s}}{\sqrt{\tilde{\tau}_d s}} \tag{1.41}$$

with an apparent diffusion time constant $\tilde{\tau}_d = \dfrac{\epsilon_p \ell^2}{\tilde{D}_e}$.

The mass balance for the detailed diffusion-convection model is:

$$D_e \frac{\partial^2 c}{\partial x^2} - v_0 \frac{\partial c}{\partial x} = \epsilon_p \frac{\partial c}{\partial t} \tag{1.42}$$

Moreover, the particle transfer function is:

$$g_p(s) = \frac{(e^{2r_2} - 1)(e^{2r_1} - 1)}{e^{2r_2} - e^{2r_1}} \frac{\sqrt{(\lambda/2)^2 + \tau_d s}}{\tau_d s} \tag{1.43}$$

with

$$r_{1,2} = \frac{\lambda}{2} \pm \sqrt{(\lambda/2)^2 + \tau_d s} \tag{1.44}$$

and $\tau_d = \dfrac{\epsilon_p \ell^2}{D_e}$.

Model equivalence leads to Equation 1.38, where the enhancement factor for pore diffusivity due to convection is expressed by Equation 1.39 with

$$f(\lambda) = \frac{3}{\lambda} \left(\frac{1}{\tanh \lambda} - \frac{1}{\lambda} \right) \tag{1.45}$$

One practical application of this concept is in the separation of proteins by HPLC. The pore velocity can be estimated from the equality between bed pressure drop relative to the bed length and particle pressure drop assuming that Darcy's law is valid; the result is $v_0 = au_0$ where a is the ratio of particle and bed permeabilities.

The Van Deemter equation [37] for conventional packings is:

$$HETP = A + \frac{B}{u} + \frac{2}{3} \frac{\epsilon_p(1 - \epsilon_b)b^2}{\left(\epsilon_b + \epsilon_p(1 - \epsilon_b)b\right)^2} \tau_d u \qquad (1.46)$$

where $HETP$ is the height equivalent to a theoretical plate, ϵ_p is the particle porosity, ϵ_b is the interparticle porosity and $b = 1 + \frac{1-\epsilon_p}{\epsilon_p} m$ with the slope of the equilibrium isotherm m; or

$$HETP = A + \frac{B}{u} + Cu \qquad (1.47)$$

For large-pore particles Rodrigues [38, 39] derived an extension of the Van Deemter equation:

$$HETP = A + \frac{B}{u} + Cf(\lambda)u \qquad (1.48)$$

At low velocities and low intraparticle Peclet number $f(\lambda) = 1$ and the Rodrigues equation (Equation 1.48) reduces to the Van Deemter equation (Equation 1.47); at high velocities $f(\lambda) = 3/\lambda$ and HETP tends to a plateau which does not depend on the solute diffusivity but only on the particle permeability and pressure gradient (convection-controlled limit); we can work faster without losing column efficiency (see Figure 1.10).

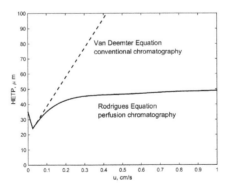

FIGURE 1.10
HETP versus superficial velocity for conventional chromatography (Van Deemter equation) and perfusion chromatography (Rodrigues equation).

1.5.2 CO_2 capture by adsorption

CO_2 capture is a challenging problem for society in relation to the greenhouse effect on climate change. Several technologies are involved (absorption with amines, membranes, ionic liquids, adsorption). The capture of CO_2 by adsorption processes includes pressure swing adsorption (PSA), vacuum swing adsorption (VPSA), or temperature swing adsorption (TSA) processes [40, 41, 42]

All these technologies involve cyclic adsorption processes and the basic element of such processes is the fixed-bed adsorption column. Each column will be subjected to various steps of the cycle: pressurization, feed, purge, blowdown, and equilibration between columns is also common practice. For a typical flue gas represented by a mixture of CO_2 and nitrogen the proper modeling of fixed-bed adsorption in some adsorbent (13X zeolite, 5A zeolite, or MOF-metal organic framework) requires:

Species mass balance

$$\epsilon_c \frac{\partial c_i}{\partial t} = \epsilon_c \frac{\partial}{\partial z}\left(c_t D_{ax}\frac{\partial y_i}{\partial z}\right) - \frac{\partial (uc_i)}{\partial z} - (1 - \epsilon_c)a'k_{fi}\frac{c_i - \bar{c}_i^p}{1 + Bi_i} \qquad (1.49)$$

where $c_t = \sum_{i=1}^{n} c_i$ and $Bi_i = \frac{R_p k_{fi}}{5\epsilon_p D_{p,i}}$.

Momentum balance (Ergun equation)

$$\frac{\partial P}{\partial z} = -\frac{150\mu(1-\epsilon_c)^2}{\epsilon_c^3 d_p^2}u + \frac{1.75(1-\epsilon_c)\rho}{\epsilon_c^3 d_p}|u|u \qquad (1.50)$$

Mass balance for the macropores (LDF model)

$$\epsilon_p \frac{\partial \bar{c}_i^p}{\partial t} + \rho_p \frac{\partial <\bar{q}_i>}{\partial t} = \epsilon_p \frac{15 D_{p,i}}{R_p^2}\frac{Bi_i}{1 + Bi_i}(c_i - \bar{c}_i^p) \qquad (1.51)$$

where $\bar{c}_i^p$ is the average fluid concentration of species in macropores (over the particle.)

Mass balance for the micropores

$$\frac{\partial <\bar{q}_i>}{\partial t} = \frac{15 D_{\mu,i}}{r_c^2}(<q_i^*> - <\bar{q}_i>) \qquad (1.52)$$

where $<\bar{q}_i>$ is the average adsorbed concentration of species in micropores (over the particle) and $<\bar{q}_i^*>$ is the average concentration over the particle of the adsorbed concentration at the entrance of the micropores.

Energy balance for the gas phase

$$\epsilon_c C_t C_v \frac{\partial T_g}{\partial t} = \frac{\partial}{\partial z}\left(\lambda \frac{\partial T_g}{\partial z}\right) - u c_t c_p \frac{\partial T_g}{\partial z} + \epsilon_c R T_g \frac{\partial c_t}{\partial t} -$$

$$- (1 - \epsilon_c) a' h_f (T_g - T_S) - \frac{2 h_w}{R_w}(T_g - T_w) \qquad (1.53)$$

Energy balance for the solid phase

$$(1 - \epsilon_c)\left(\epsilon_p \sum_{i=1}^{n} \bar{c}_i^p c_{vi} + \rho_p \sum_{i=1}^{n} <\bar{q}_i> c_{v,ads,i} + \rho_p c_{pS}\right)\frac{\partial T_S}{\partial t} =$$

$$= (1 - \epsilon_c)\epsilon_p R T_S \frac{\partial c_i}{\partial t} + \rho_b \sum_{i=1}^{n}(-\Delta H_i)\frac{\partial <\bar{q}_i>}{\partial t} +$$

$$+ (1 - \epsilon_c) a' h_f (T_g - T_S) \qquad (1.54)$$

where R is the ideal gas constant, c_p is the heat capacity at constant pressure process, and c_v is the heat capacity at constant pressure volume.

Energy balance for the column wall

$$\rho_w c_{pw} \frac{\partial T_w}{\partial t} = \alpha_w h_w (T_g - T_w) - \alpha_{wl} U (T_w - T_\infty) \qquad (1.55)$$

where U is the wall heat transfer coefficient and

$$\alpha_w = \frac{D_w}{e(D_w + e)}$$

$$\alpha_{wl} = \frac{D_w + 2e}{e(D_w + e)}$$

Adsorption equilibrium isotherm

$$\frac{q_i^*}{q_{max,i}} = K_i P y_i \left[1 - \sum_i \left(\frac{q_i^*}{q_{max,i}}\right)\right]^{a_i} \qquad (1.56)$$

$$K_i = K_i^0 \exp\left(\frac{-\Delta H_i}{R T_S}\right) \qquad (1.57)$$

Constitutive law (ideal gas)

$$P = c_t R T_g \qquad (1.58)$$

with adequate boundary and initial conditions for each step.

The solution of the model for a bed initially filled with nitrogen at the temperature of the feed allows the prediction of $u(x,t)$, $T(x,t)$, $P(x,t)$, and $c_i(x,t)$. As an example the evolution of the concentration front of CO_2 along the fixed-bed is shown in Figure 1.11 at times $0, 25, 50, 100$, and $200s$.

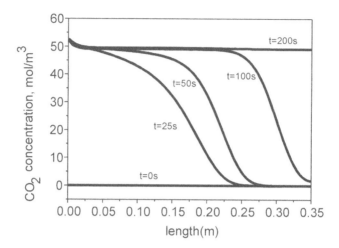

FIGURE 1.11
CO_2 concentration profiles at various times in a fixed-bed adsorber.

1.6   Conclusions

Modeling is the fingerprint of a generation of chemical engineers associated with the Second Paradigm of Chemical Engineering. Simulation tasks can be simplified with the availability of friendly software. However, one must be careful with the validation of results from numerical simulation. In many ChE problems there are moving fronts (temperature, concentration) in fixed bed processes or even at the particle scale. In such situations it is appealing to apply the moving finite elements method (MFEM) with an adaptive mesh following the moving fronts as will be discussed in the following chapters of the book.

Bibliography

[1] James Wei. New horizons in reaction engineering. *Chemical Engineering Science*, 45(8):1947–1952, 1990.

[2] National Research Council Board on Chemical Sciences and Technology. *Frontiers in Chemical Engineering: Research Needs and Opportunities.* National Academy Press, Washington D.C., 1988.

[3] W. H. Walker, W. K. Lewis, and W. H. McAdams. *The Principles of Chemical Engineering.* New York: McGraw-Hill, 1923.

[4] R. Byron Bird, Warren E. Stewart, and Edwin N. Lightfoot. *Transport Phenomena.* New York: John Wiley, 2002.

[5] J. C. Maxwell. *The Scientific Papers of James Clerk Maxwell.* New York: Dover Pubs, 1952.

[6] K. Westerterp, A. Kronberg, A. Benneker, and V. Dil'man. Wave concept in the theory of hydrodynamical dispersion — A Maxwellian type approach. *Trans. Instn. Chem. Engrs. Part A*, 74:994–952, 1996.

[7] Jing Liu, Xu Chen, and L.X. Xu. New thermal wave aspects on burn evaluation of skin subjected to instantaneous heating. *IEEE Transactions on Biomedical Engineering*, 46(4):420–428, April 1999.

[8] S. Bruin. *Product-Driven Process Engineering.* Inaugural Lecture at the TU Eindhoven, 2004.

[9] E. Cussler and G. Moggridge. *Chemical Product Design.* Cambridge University Press, 2011.

[10] K. Ulrich and S. Eppinger. *Product Design and Development.* McGraw Hill, 2008.

[11] J. Wesselingh, S. Kill, and M. Vigild. *Design and Development of Biological, Chemical, Food and Pharmaceutical Products.* Wiley, 2007.

[12] J. Wei. *Product Engineering. Molecular Structure and Properties.* Oxford University Press, 2007.

[13] A. E. Rodrigues. Adsorption. What else? *Materiales en Adsorción y Catálisis*, 4:16–26, 2012.

[14] P. Le Goff. *Cours de Cinétique Physique et Chimique.* ENSIC, 1970.

[15] R. Aris. *Mathematical Modelling Techniques.* New York: Dover Pubs, 1994.

[16] Shinnar R. Chemical reactor modeling — The desirable and the achievable. In *Chemical Reaction Engineering Reviews*, chapter 2, pages 1–36. ACS symposium series, 1978.

[17] D. Himmelblau and K. Bischoff. *Process Analysis and Simulation.* Wiley, 1968.

[18] Octave Levenspiel. Modeling in chemical engineering. *Chemical Engineering Science*, 57:4691 – 4696, 2002. Festschrift in Honour of Dr Winn van Swaaij.

[19] K. Denbigh. *The Thermodynamics of the Steady State*. Methuen, London, 1951.

[20] R. Taylor and R. Krishna. *Multicomponent Mass Transfer*. Wiley, 1993.

[21] J. Wesselingh and R. Krishna. *Mass Transfer*. Ellis Horwood, 1990.

[22] A. E. Rodrigues. Percolation theory I-basic principles and II-modeling and design of percolation columns. In J. M. Calo and E. J. Henley, editors, *Chromatography, Percolation, Adsorption and Gas Adsorption*, volume 5 of *AIChEMI Series B: Stagewise and Mass Transfer Operations*, pages 7–24. AIChE, 1984.

[23] A. E. Rodrigues and Mirjana Minceva. Modelling and simulation in chemical engineering: Tools for process innovation. *Computers and Chemical Engineering*, 29(6):1167–1183, 2005.

[24] A. E. Rodrigues. Modeling and simulation in chemical engineering. In A. E. Rodrigues, Paula de Oliveira, José Almiro Castro, José Augusto Ferreira, and Maria do Carmo Coimbra, editors, *Modeling and Simulation in Chemical Engineering*. CIM, 2003.

[25] E. Glueckauf. Theory of chromatography. part 10-formulae for diffusion into spheres and their applicability to chromatography. *Transactions of Faraday Society*, 51:1540–1551, 1955.

[26] Theodore Vermeulen. Separation by adsorption methods. In Thomas B. Drew and John W. Hoopes, editors, *Advances in Chemical Engineering*, volume 2, pages 147–208. Academic Press, 1958.

[27] P. Weisz and D. Prater. Interpretation of measurements in experimental catalysis. *Advance in Catalysis*, 6:147–205, 1954.

[28] G. Damkholer. Ubertemperatur in kontaktkornern. *Zeitschrift fur Physikalische Chemie*, 193:16–28, 1943.

[29] D. De Vault. The theory of chromatography. *Journal of American Chemical Society*, 65:532–540, 1943.

[30] A. Nir and L. Pismen. Simultaneous intraparticle forced convection, diffusion and reaction in a porous catalyst. *Chemical Engineering Science*, 32(1):35–41, 1977.

[31] A. Wheeler. Reaction rates and selectivity in catalyst pores. *Advances in Catalysis*, 3:249–327, 1951.

[32] P. V. Danckwerts. Local residence-times in continuous-flow systems. *Chemical Engineering Science*, 9(1):78–79, 1953.

[33] E. Brunier, A. Zoulalian, G. Antonini, and A. Rodrigues. Residence time distributions in laminar flow through reservoirs from momentum and mass transport equations. *Institution of Chemical Engineers Symposium Series*, 87:439–445, 1984.

[34] L. M. Madeira, M. Alves, and A.E. Rodrigues. Teaching non-ideal reactors with cfd tools. *Chemical Engineering Education*, 38:154–160, 2004.

[35] N. B. Afeyan, N. F. Gordon, I. Mazsaroff, L. Varady, S. P. Fulton, Y. B. Yang, and F. E. Regnier. Flow-through particles for the high-performance liquid chromatographic separation of biomolecules: perfusion chromatography. *Journal of Chromatography A*, 519(1):1–29, 1990.

[36] A. E. Rodrigues, B. Ahn, and Zoulalian. Intraparticle forced convection effect in catalyst diffusivity measurements and reactor design. *AIChE. J*, 28:925–930, 1982.

[37] J. Van Deemter, F. Zuiderweg, and A. Klinkenberg. Longitudinal diffusion and resistance to mass transfer as causes of nonideality in chromatography. *Chemical Engineering Science*, 5:271–289, 1956.

[38] A. E. Rodrigues, Lu Zuping, and J. M. Loureiro. Residence time distribution of inert and linearly adsorbed species in a fixed bed containing large-pore supports: applications in separation engineering. *Chemical Engineering Science*, 46(11):2765–2773, 1991.

[39] A. E. Rodrigues. An extended Van Deemter equation (Rodrigues equation) for performing chromatographic processes using large-pore, permeable packings. *LC-GC*, 6(1):20–29, 1993.

[40] Zhen Liu, Lu Wang, Xiangming Kong, Ping Li, Jianguo Yu, and Alírio E. Rodrigues. Onsite CO2 capture from flue gas by an adsorption process in a coal-fired power plant. *Industrial and Engineering Chemistry Research*, 51(21):7355–7363, 2012.

[41] Lu Wang, Zhen Liu, Ping Li, Jianguo Yu, and Alírio E. Rodrigues. Experimental and modeling investigation on post-combustion carbon dioxide capture using zeolite 13X-APG by hybrid VTSA process. *Chemical Engineering Journal*, 197:151–161, 2012.

[42] Paul A. Webley and Jianming He. Fast solution-adaptive finite volume method for PSA/VSA cycle simulation; 1 single step simulation. *Computers and Chemical Engineering*, 23(11-12):1701–1712, 2000.

2

The Moving Finite Elements Method

2.1 Mathematical Models

Mathematical models comprise a set of equations whose solution describes the behavior of the related physical problem. Mathematical models that are described by differential equations that include functions of several variables and their partial derivatives are used to formulate problems in science and engineering. The realization that entirely different physical problems may have identical or similar mathematical formulations is a stimulus for new advances in science. Mathematical models are created taking into account the phenomena present in the process and by applying laws and theories that describe the mass conservation, momentum, and energy [1]. In general, a mathematical model is a simplified description of real problems. However, building mathematical models requires a thorough understanding of the problem to be solved. This demand produces nonlinear models, and the significant progress concerning the classic solution of partial differential equations (PDE) made over the last two centuries only solves a small part of the simplest real problems. Finding their solutions by analytical procedures such as Laplace and Fourier methods is either impossible or impracticable since usually analytic solution are in the form of an integral or a series, requiring numerical evaluations. Thus most applications require numerical solutions, and efficient numerical algorithms are needed to represent an approximation to the equations describing the mathematical model. For that reason, it remains highly desirable to know some aspects of the behavior of the solution and the classical conventional classification of partial differential equations is extremely useful. For second-order linear partial differential equations, the theory defines three distinct kinds of partial differential equations: elliptic, hyperbolic and parabolic. We shall adopt a less rigorous classification based on the general properties of their solutions. For an elliptic equation, we can think of the Laplace equation. An example of a hyperbolic equation is the wave equation, and the diffusion equation is representative of a parabolic equation [2]. Many other problems have solutions exhibiting behaviors that change from one region to another [3].

Partial differential equations to describe natural phenomena require four independent variables, the three-dimensional space, and time. Auxiliary conditions must be specified to ensure that the problem has a unique solution, as well as a correct physical meaning. In that case, we will say that the problem is well-posed. The type of PDE defines which initial and boundary conditions must be imposed to have a well-posed problem. Initial conditions are needed to establish the value of the dependent variables at the initial value of independent variable time. Usually, this condition refers to $t_0 = 0$. As the name suggests, boundary conditions indicate the behavior of the solution at the boundary of the space domain. Both initial and boundary conditions define how the problem starts. Different kinds of boundary conditions may be considered. Dirichlet boundary condition is a condition where the value of the dependent variable is prescribed on the boundary. Neumann boundary condition is a condition where the value of the space derivative of the dependent variable is known on the boundary. A third type or Robin boundary condition is a combination of the Dirichlet and the Neumann boundary condition.

2.1.1 Linear convection–diffusion model

Let us start with the standard linear mathematical convection–diffusion problem. This model deals with the evolution of a particular quantity such as the concentration of a chemical species in a flow medium under the influence of two processes, convection and diffusion. Convection is associated with the flow fluid transport and refers to the movement of species within a medium determined by a known velocity field. In the literature concerning transport phenomena, convection is also referred to as advection. Suppose that $u = u(x,t)$ is a concentration of a certain species. If the species is transported with velocity ν, then the mass conservation law implies that the concentration satisfies the convection equation

$$\frac{\partial u}{\partial t} + \frac{\partial (\nu u)}{\partial x} = 0 \qquad (2.1)$$

For ease of notation, we frequently omit the explicit dependence of u and its derivatives, on x and t.

Diffusion is another fundamental transport process associated with the movement from an area of high concentration to an area of low concentration, with an equilibrium state of uniform concentration. In addition to convection, if we consider the effect of the change of u caused by the gradients in the solution with diffusion coefficient d the corresponding equation is

$$\frac{\partial u}{\partial t} + \frac{\partial (\nu u)}{\partial x} = \frac{\partial}{\partial x}\left(d\frac{\partial u}{\partial x}\right) \qquad (2.2)$$

where $x \in \Omega$, Ω is an interval of $\mathbb{R}$, $t > 0$ and the equation is subject to given initial and boundary conditions. Even assuming that the given coefficients ν and d are independent of the concentration, there are very few analytical solutions to the linear convection–diffusion equation with variable coefficients. The analytical solutions found in the literature deal with particular cases of variable constants [4, 5, 6]. Zamani and Bombardelli [6] give an appealing table presenting particular models for which the analytical solutions can be found in the literature.

2.1.2 Nonlinear convection–diffusion–reaction model

In many applications, models must take into consideration the local change in the dependent variable due to sources and chemical reactions modeled by the nonlinear reaction equation

$$\frac{\partial u}{\partial t} = S\left(x, t, u\right) \tag{2.3}$$

Combining the convection–diffusion equation 2.2 with reaction equation 2.3, the convection–diffusion–reaction equation may be written as

$$\frac{\partial u}{\partial t} + \frac{\partial (\nu u)}{\partial x} = \frac{\partial}{\partial x}\left(d\frac{\partial u}{\partial x}\right) + S\left(x, t, u\right) \tag{2.4}$$

A more realistic model should consider more than one unknown variable. For example, chemical reactions occur between two or more species, or adsorption process with two or more components, which yields a system of equations [7], [8], [9]. So, instead of one single dependent variable we will consider a vector of dependent variables, $\mathbf{u}(x,t) = [u_1(x,t), u_2(x,t), ..., u_n(x,t)]^T$, the superscript T denotes the vector or matrix transpose operation. The independent variables are $x \in \Omega = [a, b]$, and time $t \geq 0$. Assuming, in addition, variable coefficients in Equation 2.4, we define a more general implicit system of n dependent variables and n equations of the form

$$\sum_{j=1}^{n} g_{m,j}\left(x, t, \mathbf{u}\right) \frac{\partial u_j}{\partial t} = \sum_{j=1}^{n} f_{m,j}\left(x, t, \mathbf{u}, \frac{\partial \mathbf{u}}{\partial x}\right) \frac{\partial^2 u_j}{\partial x^2}$$

$$+ h_m\left(x, t, \mathbf{u}, \frac{\partial \mathbf{u}}{\partial x}\right), m = 1, 2, ...n \tag{2.5}$$

under certain initial conditions and given Dirichlet, Neumann, or Robin boundary conditions. The functions $f_{m,j}$ and h_m depend on space and time and can also depend on the unknown solution $\mathbf{u}$ and on its space derivative $\mathbf{u}_x = \frac{\partial \mathbf{u}}{\partial x}$. The diffusive coefficient is defined by $f_{m,j}$ and h_m includes convection and reaction phenomena. To simplify the notation we use a matrix notation. Let $\mathbf{G}$ be the matrix with function $g_{m,j}$ on position G_{mj}. Each entry of $\mathbf{G}$ depends on the space, time, and on the unknown solution $\mathbf{u}$. In a similar manner let $\mathbf{F}$ be the matrix with function $f_{m,j}$ in position F_{mj} and $\mathbf{h}$ is the vector of functions $h_m, m = 1, ..., n$. With this notation the nonlinear convection–diffusion–reaction model can be expressed by the equation

$$\mathbf{G}\left(x, t, \mathbf{u}\right) \frac{\partial \mathbf{u}}{\partial t} = \mathbf{F}\left(x, t, \mathbf{u}, \mathbf{u}_x\right) \frac{\partial^2 \mathbf{u}}{\partial x^2} + \mathbf{h}\left(x, t, \mathbf{u}, \mathbf{u}_x\right), x \in \Omega, \, t > 0 \tag{2.6}$$

and appropriated initial and boundary conditions. By a solution for the model 2.6, we mean a function $\mathbf{u} = \mathbf{u}(x,t)$ defined on the space-time domain, that satisfies Equation 2.6 and the initial and boundary conditions. We shall suppose that $\mathbf{u}$ has as many continuous partial derivatives as required by the PDE. If $\mathbf{G}$ is the identity matrix, then the system becomes an explicit system. Equation 2.6 is linear if it can be expressed as a linear combination of function $\mathbf{u}$ and its partial derivatives. Otherwise, the PDE is nonlinear. Boundary conditions can also be nonlinear.

Equation 2.6 is used as a mathematical model for a wide range of physical problems in all branches of engineering and science [3, 10, 11, 12, 13].

The numerical resolution of time-dependent partial differential problems implies that both space and time domains are discretized. Thus Equation 2.6 and also initial and boundary conditions are replaced by a discrete model to be solved in a computer. Various techniques are available to obtain the discrete model. Two excellent books are provided by Ames [14] and by Finlayson [15] for an overview of the numerical analysis of partial differential equations. The book by Hundsdorfer and Verwer [16] deals with numerical methods for solving this kind of equation, with a focus on the time dependency. The most widely numerical methods use finite element [17] and finite difference methods [18] for discretizing and solving partial differential equations, but alternative methods of solution are often suitable for special classes of problems [19]. The principle of finite differences methods consists in approximating the differential operator by replacing the partial derivatives in the differential equation with their finite difference approximations. For theoretical results regarding accuracy, stability, and convergence of finite differences methods for partial differential equations, the reader is referred to [20]. Rather than finding the solution at the mesh nodes the idea behind the finite element methods is to approximate the solution with suitable functions. A finite combination of basis functions that are zero except on a small part of the domain are considered. The problem of solving the differential equation is converted into a corresponding problem of minimizing a functional [21, 22]. The most usually adopted basis functions are piecewise linear functions, even though they are not the most efficient.

2.2   The MFEM in One-Dimensional (1D) Space Domain

2.2.1   MFEM formulation

It is well known that some solutions of problems arising in science and engineering present a steep front that moves through the spatial domain. The discretization of space domain with a fixed mesh can be inefficient since a large number of spatial nodes is required to capture the solution in the regions of high spatial activity, adequately. For this reason, the literature concerning adaptive moving mesh methods is growing rapidly, and some relevant books [23, 24, 25] and review articles [26, 27, 28] addressing moving mesh methods can be found.

In adaptive moving mesh methods, the model equations can be solved by a method of lines [29, 30, 31] approach: the space domain is discretized by using a fixed number of mesh points (nodes) moving in time, keeping the same mesh topology. The discretization in space leads to solving a set of simultaneous ordinary differential equations (ODE). In fact, we want that the semi-discrete model preserves as closely as possible all the critical properties of the original equation with the fewest possible number of nodes. For that purpose, adaptive mesh is essential. These methods are designed to concentrate the spatial nodes in regions of high spatial activity. Thus, the computational effort focuses on the spatial region where it is most needed.

The moving finite element, originally developed by Miller and Miller [32] and Miller [33], is a moving mesh method in which the mesh deforms continuously in time. Important variants proposed by Miller and co-workers include the gradient-weighted moving finite elements, [34, 35, 36], least squares moving finite elements method [37], stabilized moving finite elements method [38]. Besides, other alternatives have been investigated. These include the use of higher-order basis functions [39, 40, 41, 8, 42, 43, 44, 45].

Here we develop and analyze an explicit formulation of the moving finite elements method (MFEM) that differs from its original implementation. The MFEM is a finite element framework that extends the notion of a basis function allowing the nodal positions to move dynamically in time. Therefore, the MFEM computes simultaneously the approximate solution and the mesh where this solution is represented. In our MFEM formulation, we are concerned with the use of arbitrary degree piecewise polynomials basis functions. Solutions are calculated using a Galerkin approach with a piecewise high degree polynomial basis in space. Lagrange polynomials are used to define the basis functions in each finite element that are themselves time-dependent through the time dependence of the nodal position.

We consider the use of different spatial mesh for each one of the dependent variables. To avoid possible singularities penalty functions are added. Another issue to be considered is the fact that piecewise polynomials do not have enough continuity properties, and a smoothed function must be considered in a δ neighborhood of a spatial node. The integrals needed to define the MFEM equations are calculated using the smoothed function, and then the limit is taken as $\delta \to 0$.

2.2.2 Spatial discretization

The moving finite elements method is a computational approach for solving partial differential equations of the form defined by Equation 2.6 that proceeds in two separate stages. The first one is the discretization of the spatial domain by finite elements allowing the movement of nodes over time. In the second, the resulting system of ordinary differential equations is integrated in the temporal variable.

The first step to define the numerical approximation to the solution $\mathbf{u}$ of model 2.6 is to discretize the spatial domain $\Omega = [a, b]$ by a finite number of nodes to define a spatial mesh. For problems with more than one dependent variable, we choose to define a spatial mesh for each dependent variable. It is a deliberate choice, causing additional computational costs, but this procedure is implemented to produce accurate numerical solutions for a class of problems having solutions with different time scales. We shall define n spatial meshes. Let $\mathcal{M}_m$ denote the space mesh associated to function u_m and q_m the fixed number of subdomains on mesh $\mathcal{M}_m$. Each spatial mesh $\mathcal{M}_m$ is time dependent as a consequence of the time dependence of spatial nodes position $X_{m,j}(t)$, $j = 1, ..., q_m + 1$.

$$\mathcal{M}_m: \quad a \leq X_{m,1}(t) < ... < X_{m,j}(t) < ... < X_{m,q_m+1}(t) \leq b \quad (2.7)$$

In this context, subdomains $\Omega_{m,j} = [X_{m,j}, X_{m,j+1}]$ are referred to as the j-th finite element of the mesh $\mathcal{M}_m$ at time t and $Z_{m,j} = X_{m,j+1} - X_{m,j}$ denotes the length of element $\Omega_{m,j}$. To simplify the notation we shall drop the time label and thus leave implicit the time-dependence of spatial nodes and finite elements and other variables constructed from them.

Let us see how moving finite elements method defines the approximation $\mathbf{U}$ to $\mathbf{u}$. Firstly we will consider the procedure, element by element, to define each component of $\mathbf{U}$ and its derivatives on a finite element. Then we will define the approximate solution over the spatial domain Ω.

2.2.3   Element-by-element approach

Consider $\Omega_{m,j}$, the j-th finite element of the mesh $\mathcal{M}_m$, and let U_m be the m-th component of the approximation $\mathbf{U}$. The approximate solution U_m on $\Omega_{m,j}$ is a polynomial function obtained using $p_{m,j}$ Lagrange basis functions and $p_{m,j}$ interpolation points on $\Omega_{m,j}$. Here the interior points within the interval are chosen to be the zeros of a $p_{m,j} - 2$ degree Jacobi polynomial, and the two endpoints are purposely included as interpolation points. This inclusion will enable us to ensure the continuity of the approximation U_m on Ω.

Consider the normalized domain $[0, 1]$ as a reference element. The Jacobi polynomials of degree p for weight function $x^\beta(1-x)^\alpha$ are defined by the orthogonality conditions

$$\int_0^1 x^\beta(1-x)^\alpha x^k P_p^{(\alpha,\beta)}(x)dx = 0, \ k = 0, 1, ..., p-1 \quad (2.8)$$

Let us first define the $p_{m,j} - 2$ interior points on the normalized interval $[0, 1]$. These points, $\xi_{m,j}^k$, $k = 2, \cdots, p_{m,j} - 1$, are selected to be the zeros of a Jacobi polynomial $P^{(\alpha,\beta)}_{p_{m,j}-2}$, $\alpha = \beta = 1$, on the normalized interval $0 \leq \xi \leq 1$. The endpoints are $\xi^1_{m,j} = 0$ and $\xi^{p_{m,j}}_{m,j} = 1$. The nodes $\xi^k_{m,j}$, $k = 1, \cdots, p_{m,j}$ are the normalized local nodes and define the local normalized mesh for polynomials of degree $p_{m,j} - 1$.

$$0 = \xi^1_{m,j} < ... < \xi^k_{m,j} < ... < \xi^{p_{m,j}}_{m,j} = 1 \tag{2.9}$$

Using these local nodes in the reference interval $[0, 1]$ and defining the relative positions of interpolation points in the normalized interval, the Lagrange basis polynomials are

$$\ell^i_{m,j}(\xi) = \prod_{k=1, k \neq i}^{p_{m,j}} \frac{\xi - \xi^k_{m,j}}{\xi^i_{m,j} - \xi^k_{m,j}}, \quad i = 1, ..., p_{m,j} \tag{2.10}$$

$0 \leq \xi \leq 1$. Here the index m, j that affects the nodes of the reference finite element is used because we allow the user to choose the degree of the polynomial approximation in each of the finite elements of each mesh. We first define $p_{m,j}$ nodes on the reference element and then by the change of variable

$$x = X_{m,j} + \xi \left(X_{m,j+1} - X_{m,j} \right) \tag{2.11}$$

the local mesh on $\Omega_{m,j}$ is established. This change of variable 2.11 transforms the set of the normalized local nodes $\xi^k_{m,j}$ to a set of local nodes, $s^k_{m,j}$, on the finite element $\Omega_{m,j}$

$$X_{m,j} = s^1_{m,j} < ... < s^k_{m,j} < ... < s^{p_{m,j}}_{m,j} = X_{m,j+1} \tag{2.12}$$

The transformation 2.11 also guarantees that $\ell^i_{m,j}(x) = \ell^i_{m,j}(\xi)$. For computational purposes, the use of a reference finite element, the normalized interval, allows us to reduce the number of calculations.

To illustrate the connection between the local normalized points and the local spatial nodes, consider a finite element with $p_{m,j} = 4$ local nodes. The local mesh on $[0, 1]$ is the set of 4 points, the two endpoints $\xi^1_{m,j} = 0$ and $\xi^4_{m,j} = 1$ and two interior points chosen as the zeros of the Jacobi polynomial of degree 2, $\xi^2_{m,j} = \frac{5-\sqrt{5}}{10}$ and $\xi^3_{m,j} = \frac{5+\sqrt{5}}{10}$. Figure 2.1 shows the two local meshes. The local normalized spatial mesh defined by $\xi^k_{m,j}$, $k = 1, \cdots, 4$ and the corresponding local spatial mesh defined by $s^k_{m,j}$, $k = 1, \cdots, 4$ on the finite element $\Omega_{m,j}$. The relative positions of the local nodes are the same for all finite elements having two local interior nodes.

With these preliminaries, we are prepared to define the restriction of U_m to the finite element $\Omega_{m,j}$. In what follows, $U_{m,j}(x, t)$ represents the restriction of U_m to $\Omega_{m,j}$ and $U^i_{m,j}$ is assumed to be the value of U_m at the i-th local spatial node on $\Omega_{m,j}$, i.e., $U^i_{m,j} = U_{m,j}(s^i_{m,j}, t)$.

Thus $U_{m,j}(x, t)$ is written in the form

$$U_{m,j}(x, t) = \sum_{i=1}^{p_{m,j}} \ell^i_{m,j}(x) U^i_{m,j}(t) \tag{2.13}$$

where $x \in \Omega_{m,j}$.

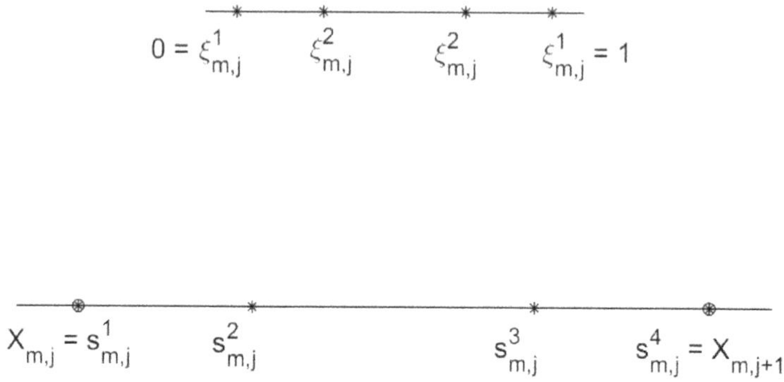

FIGURE 2.1

Local normalized spatial mesh on reference element and local spatial mesh on $\Omega_{m,j}$.

The spatial derivatives of $U_{m,j}(t)$ on the finite element $\Omega_{m,j}$ are also polynomials, and will be defined using Lagrange interpolation and the local spatial nodes $s_{m,j}^i$,

$$\frac{\partial U_{m,j}}{\partial x}(x,t) = \sum_{i=1}^{p_{m,j}} \ell_{m,j}^i(x) \frac{\partial U_{m,j}}{\partial x}\left(s_{m,j}^i\right) \qquad (2.14)$$

$$\frac{\partial^2 U_{m,j}}{\partial x^2}(x,t) = \sum_{i=1}^{p_{m,j}} \ell_{m,j}^i(x) \frac{\partial^2 U_{m,j}}{\partial x^2}\left(s_{m,j}^i\right) \qquad (2.15)$$

where $x \in \Omega_{m,j}$, and $t > 0$.

Introducing the normalized local coordinates and defining the quantities

$$A_{m,j}^k(\xi^i) = \frac{d\ell_{m,j}^k(x)}{d\xi}\Big|_{\xi^i}$$
$$B_{m,j}^k(\xi^i) = \frac{d^2\ell_{m,j}^k(x)}{d\xi^2}\Big|_{\xi^i} \qquad (2.16)$$

then the first spatial derivative and the second spatial derivatives computed at the local spatial nodes $s_{m,j}^i, i = 1, 2, ..., p_{m,j}$, appearing in 2.14 and 2.15, are of the form

$$\frac{\partial U_{m,j}}{\partial x}\left(s_{m,j}^i\right) = (Z_{m,j})^{-1} \sum_{k=1}^{p_{m,j}} A_{m,j}^k(\xi^i) U_{m,j}^k \qquad (2.17)$$

$$\frac{\partial^2 U_{m,j}}{\partial x^2}\left(s_{m,j}^i\right) = (Z_{m,j})^{-2} \sum_{k=1}^{p_{m,j}} B_{m,j}^k(\xi^i) U_{m,j}^k \qquad (2.18)$$

So far we have attempted to define the restriction of solution U_m and its spatial derivatives on the finite element $\Omega_{m,j}$. Approximate solutions are constructed element by element and then the global solution must be defined.

2.2.4 Global approach

Consider the component function U_m defined on $\mathcal{M}_m$. This global function can be written as the sum of all local element contributions. U_m is taken as a polynomial piecewise function such as the restriction of U_m to $\Omega_{m,j}, j = 1, \cdots, q_m$ is given by Equation 2.13. For that purpose let us define the ordered set of all spatial nodes and all interior local nodes associated to the space mesh $\mathcal{M}_m$,

$$\Delta_m = \bigcup_{j=1}^{q_m} \{s^k_{m,j}, \ k = 1, ..., p_{m,j}\} \qquad (2.19)$$

The set Δ_m is the ordered set of the global nodes associated to the space mesh $\mathcal{M}_m$. A global node is noted by $\Delta_{m,i}, i = 1, \cdots, D_m$. Here D_m is the number of elements in Δ_m.

Figure 2.2 shows an example of a $\mathcal{M}_m$ spatial mesh with 3 finite elements and the global nodes corresponding to a piecewise polynomial of degree 2 in the first element, degree 3 in the second element, and a linear approximation in the third element. A global spatial node can be either a spatial node or an interior local node. In this example there are 4 spatial nodes and 3 interior local nodes so the global mesh has 7 global nodes, $D_m = 7$.

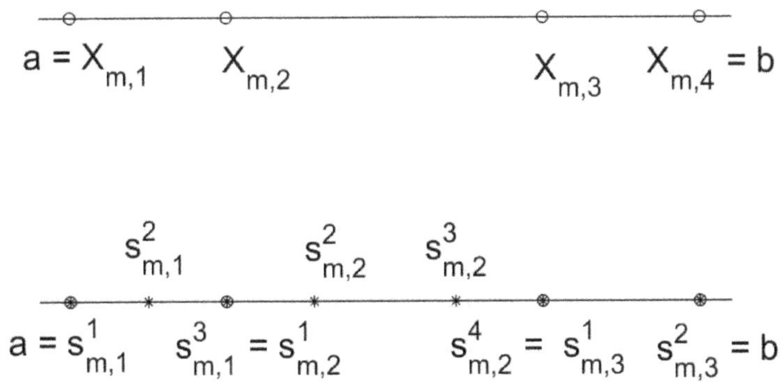

FIGURE 2.2
Global nodes: Spatial nodes and local interior nodes of $\mathcal{M}_m$.

The global approximation U_m to u_m on Ω can be written as

$$U_m(x,t) = \sum_{i=1}^{D_m} \Phi_{m,i}(x) U_m(\Delta_{m,i}) \tag{2.20}$$

where $U_m(\Delta_{m,i})$ is the value of U_m at the global node $\Delta_{m,i}$ and time t. $\Phi_{m,i}$ is the global basis function verifying

$$\Phi_{m,i}(\Delta_{m,l}) = \begin{cases} 0, & i \neq l \\ 1, & i = l \end{cases} \tag{2.21}$$

If $\Delta_{m,i}$ is an interior local spatial node then $\Delta_{m,i} = s_{m,j}^k$ for some $j = 1, \cdots, q_m$ and $k = 2, \cdots, p_{m,j}$. Then the corresponding function $\Phi_{m,i}$ is

$$\Phi_{m,i}(x) = \begin{cases} \ell_{m,j}^k(x), & x \in \Omega_{m,j} \\ 0, & \text{otherwise} \end{cases} \tag{2.22}$$

If $\Delta_{m,i}$ is a spatial node then $\Delta_{m,\,i} = X_{m,j+1}$ for some $j = 2, \cdots, q_m - 1$. Then the corresponding function $\Phi_{m,i}$ is

$$\Phi_{m,i}(x) = \begin{cases} \ell_{m,j}^{p_{m,j}}(x), & x \in \Omega_{m,j} \\ \ell_{m,j+1}^{1}(x), & x \in \Omega_{m,j+1} \\ 0, & \text{otherwise} \end{cases} \tag{2.23}$$

The piecewise polynomial function U_m defined by 2.20 depends on the position of the spatial nodes $X_{m,j}$, $j = 1, \cdots, q_{m,j}$, and also depends on nodal amplitudes $U_{m,j}^i$, $j = 1, \cdots, q_{m,j}$, $i = 1, \cdots, p_{m,j}$, which represents the value of U_m at an interpolation point of a finite element of spatial mesh $\mathcal{M}_m$. Both $X_{m,j}$ and $U_{m,j}^i$ are time dependent and therefore the time derivative of U_m will be of the form

$$\begin{aligned} \frac{\partial U_m}{\partial t} = & \sum_{j=1}^{q_m} \left[\sum_{k=2}^{p_{m,j}-1} \frac{\partial U_m}{\partial U_{m,j}^k} \frac{dU_{m,j}^k}{dt} \right] \\ & + \sum_{j=1}^{q_m-1} \frac{\partial U_m}{\partial U_{m,j}^{p_{m,j}}} \frac{dU_{m,j}^{p_{m,j}}}{dt} + \sum_{j=1}^{q_m-1} \frac{\partial U_m}{\partial X_{m,j+1}} \frac{dX_{m,j+1}}{dt} \\ & + \left(\frac{\partial U_m}{\partial U_{m,\,1}^1} \frac{dU_{m,1}^1}{dt} \right) + \left(\frac{\partial U_m}{\partial U_{m,q_m}^{p_{m,q_m}}} \frac{dU_{m,q_m}^{p_{m,q_m}}}{dt} \right) \end{aligned} \tag{2.24}$$

To determine $\dfrac{\partial U_m}{\partial t}$ it requires the evaluation of the U_m derivatives with respect to mesh deformation,

$$\frac{\partial U_m}{\partial U_{m,j}^k} \tag{2.25}$$

and

$$\frac{\partial U_m}{\partial X_{m,j+1}} \tag{2.26}$$

The derivative 2.25 is obtained by noting that $U_m(\Delta_{m,i}) = U^k_{m,j}$ for some j and k, $j = 1, \cdots, q_m$ and $k = 1, \cdots, p_{m,j}$. From 2.20 we can establish that

$$\frac{\partial U_m}{\partial U^k_{m,j}} = \Phi_{m,i}(x) \tag{2.27}$$

Therefore, if the global node $s^k_{m,j}$ is an interior local node then from 2.22 and 2.27 we write the derivatives appearing in 2.24 as

$$\frac{\partial U_m}{\partial U^k_{m,j}} = \begin{cases} \ell^k_{m,j}(x), & x \in \Omega_{m,j} \\ 0, & \text{otherwise} \end{cases} \tag{2.28}$$

In the case that $s^k_{m,j}$ is a spatial node, then from 2.23 and 2.27 we have

$$\frac{\partial U_m}{\partial U^{p_{m,j}}_{m,j}} = \begin{cases} \ell^{p_{m,j}}_{m,j}(x), & x \in \Omega_{m,j} \\ \ell^1_{m,j+1}(x), & x \in \Omega_{m,j+1} \\ 0, & \text{otherwise} \end{cases} \tag{2.29}$$

Concerning the derivative defined by Equation 2.26, let us define the variation of a function defined on an interval with respect to the variations of one of the endpoints [46, 41]. Let $f : [\bar{a}, \bar{b}] \to \mathbb{R}$ and suppose that the right end changes from $\bar{b}$ to $\bar{b} + h$. We define

$$\frac{df}{d\bar{b}} = \lim_{h \to 0} \frac{f(g(0)) - f(g(h))}{h} \tag{2.30}$$

where $g(h) = \bar{a} + \frac{\bar{b}-\bar{a}+h}{\bar{b}-\bar{a}}(x - \bar{a})$. Note that the transformation $c(x) = \bar{a} + \frac{\bar{b}+h-\bar{a}}{\bar{b}-\bar{a}}(x - \bar{a})$ maps $[\bar{a}, \bar{b}]$ into $[\bar{a}, \bar{b} + h]$. Then

$$\frac{df}{d\bar{b}} = \lim_{h \to 0} -\frac{df}{dg}\frac{dg}{dh} \tag{2.31}$$

and therefore

$$\frac{df}{d\bar{b}} = -\frac{df}{dx}\frac{x - \bar{a}}{\bar{b} - \bar{a}} \tag{2.32}$$

In a similar way, supposing that the left end changes from $\bar{a}$ to $\bar{a} + h$, by using the auxiliary function $g(h) = \bar{a} + h + \frac{\bar{b}-\bar{a}-h}{\bar{b}-\bar{a}}(x - \bar{a})$, it can be shown that

$$\frac{df}{d\bar{a}} = -\frac{df}{dx}\left(1 - \frac{x - \bar{a}}{\bar{b} - \bar{a}}\right) \tag{2.33}$$

These results provide a relationship between the nonstandard derivative of U_m with respect to the nodal position $X_{m,j+1}$ and the conventional spatial derivative of the restriction of U_m to some $\Omega_{m,j}$ allowing the calculation of the derivatives $\dfrac{\partial U_m}{\partial X_{m,j+1}}$ appearing in Equation 2.24. If fact, assuming a small variation in $X_{m,j+1}$ and keeping all other spatial nodes in their positions then the global solution varies only in the finite elements $\Omega_{m,j}$ and $\Omega_{m,j+1}$. For all $x \in \Omega_{m,j}$, Equation 2.32 gives

$$\frac{\partial U_m}{\partial X_{m,j+1}} = -\frac{\partial U_{m,j}}{\partial x}\frac{x - X_{m,j}}{Z_{m,j}} \tag{2.34}$$

For all $x \in \Omega_{m,j+1}$, Equation 2.33 tells us that

$$\frac{\partial U_m}{\partial X_{m,j+1}} = -\frac{\partial U_{m,j+1}}{\partial x}\left(1 - \frac{x - X_{m,j+1}}{Z_{m,j+1}}\right) \tag{2.35}$$

and in all other finite elements, $\dfrac{\partial U_m}{\partial X_{m,j+1}} = 0$.

On the other hand, the derivatives 2.26 appearing in Equation 2.34 and Equation 2.35 are polynomials and can be expressed by using Lagrange polynomials with $p_{m,j}$ and $p_{m,j+1}$ local nodes, respectively. This leads to

$$\frac{\partial U_m}{\partial X_{m,j+1}} = \begin{cases} -\displaystyle\sum_{k=1}^{p_{m,j}} \ell_{m,j}^k \xi_{m,j}^k \left.\frac{\partial U_{m,j}}{\partial x}\right|_{s_{m,j}^k}, & x \in \Omega_{m,j} \\[4mm] -\displaystyle\sum_{k=1}^{p_{m,j+1}} \ell_{m,j+1}^k \left(1 - \xi_{m,j+1}^k\right) \left.\frac{\partial U_{m,j+1}}{\partial x}\right|_{s_{m,j+1}^k}, & x \in \Omega_{m,j+1} \\[4mm] 0, & \text{otherwise} \end{cases} \tag{2.36}$$

Equation 2.36 allows to write the derivatives of U_m with respect to node movements in terms of conventional spatial derivatives which completes the definition of $\dfrac{\partial U_m}{\partial t}$.

2.2.5   Residual minimization

Having derived an expression for $\dfrac{\partial U_m}{\partial t}$ we now follow the procedure to find the approximate solution. In the Equation 2.24 defining $\dfrac{\partial U_m}{\partial t}$ there are two sets of unknowns in this equation, more specifically $\dfrac{dU_{m,j}^k}{dt}$ and $\dfrac{dX_{m,j}}{dt}$. The semi-discrete equations are obtained by minimizing

$$\sum_{m=1}^{n} \int_a^b (R_m)^2 dx \tag{2.37}$$

over all possible choices of $\dfrac{\partial \mathbf{U}}{\partial t}$, yielding a system of ordinary differential equations. Here R_m is the residual associated with the *m-th* partial differential equation.

$$R_m = \sum_{l=1}^{n} g_{m,l}\left(x, t, \mathbf{U}\right)\frac{\partial U_l}{\partial t} - \sum_{l=1}^{n}\left[f_{m,l}\left(x, t, \mathbf{U}, \frac{\partial \mathbf{U}}{\partial x}\right)\frac{\partial^2 U_l}{\partial x^2}\right]$$
$$- h_m\left(x, t, \mathbf{U}, \frac{\partial \mathbf{U}}{\partial x}\right) \qquad (2.38)$$

This procedure generates a system of ordinary differential equations allowing the determination of nodal amplitudes, $U_{m,j}^k$, and the nodal positions, $X_{m,j}$.

The MFEM will be said to degenerate if the mass matrix of the resulting system of ordinary differential equations is singular or if one element of the meshes does not have positive measure. The second condition occurs whenever there exists a j such that $X_{m,j+1} - X_{m,j}$ is not positive at a certain time t. The mesh will be said to be a degenerate mesh.

To overcoming the problem of a potentially singular MFEM mass matrix a suitable chosen penalty term is added to the objective function $\mathcal{T}$. This idea was presented in [32, 33] by Miller. Following the same idea the penalty term takes the form

$$\sum_{m=1}^{n}\sum_{j=1}^{q_m}\left(\epsilon_{m,j}\frac{dZ_{m,j}}{dt} - S_{m,j}\right)^2 \qquad (2.39)$$

where $S_{m,j}$ and $\epsilon_{m,j}$ are the penalty functions. These functions depend on six small positive constants, $c_i^{m,j}$, chosen by the user. The most general form of penalty functions appearing in the literature [32, 33, 47, 41] is

$$S_{m,j} = \left(\frac{c_1^{m,j}}{Z_{m,j}-c_5^{m,j}} - c_2^{m,j}(Z_{m,j} - c_5^{m,j})\right)\left(1 + \frac{c_6^{m,j}}{Z_{m,j}-c_5^{m,j}}\right)^2$$
$$(2.40)$$
$$\epsilon_{m,j} = \left(\frac{c_3^{m,j}}{Z_{m,j}-c_5^{m,j}} + c_4^{m,j}\right)\left(1 + \frac{c_6^{m,j}}{Z_{m,j}-c_5^{m,j}}\right)^2$$

Authors refer to the penalty functions $\epsilon_{m,j}$ as the internodal viscosity, and $S_{m,j}$ as a internodal spring function. In particular, the penalty constant $c_5^{m,j}$ can be seen as a minimum length permissible for the corresponding finite element.

The chosen penalty term must modify the equations in order to affect mainly the mesh, rather than the solution which is represented on it. In place of 2.37 we shall minimize

$$\mathcal{T} = \sum_{m=1}^{n}\left[\int_a^b (R_m)^2 dx + \sum_{j=1}^{q_m}\left(\epsilon_{m,j}\frac{dZ_{m,j}}{dt} - S_{m,j}\right)^2\right] \qquad (2.41)$$

with respect to

$$\dot{U}_{m,j}^k = \frac{dU_{m,j}^k}{dt}, \quad \begin{cases} m = 1, 2, ..., n \\ j = 1, 2, ..., q_m \\ k = 1, 2, ..., p_{m,j} \end{cases} \qquad (2.42)$$

and

$$\dot{X}_{m,j+1} = \frac{dX_{m,j+1}}{dt}, \begin{cases} m = 1, 2, ..., n \\ j = 1, 2, ..., q_m - 1 \end{cases} \tag{2.43}$$

Note that $X_{m,1}$ and X_{m,q_m+1} are the fixed boundary of the physical space domain. Thus the resulting ODE system is

$$\begin{cases} \dfrac{\partial \mathcal{T}}{\partial \dot{U}_{m,j}^k} = 0 \\ \\ \dfrac{\partial \mathcal{T}}{\partial \dot{X}_{m,j+1}} = 0 \end{cases} \tag{2.44}$$

Differentiating $\mathcal{T}$ with respect $\dot{U}_{m,j}^k$, the first equation of system 2.44 is

$$\sum_{i=1}^{n} \int_a^b R_i \frac{\partial R_i}{\partial \dot{U}_{m,j}^k} dx = 0 \tag{2.45}$$

From Equation 2.38 and Equation 2.24 the first equation of system 2.44 takes the form

$$\sum_{i=1}^{n} \int_a^b R_i g_{i,m} \frac{\partial U_m}{\partial U_{m,j}^k} dx = 0 \tag{2.46}$$

Concerning the second equation of system 2.44, differentiating $\mathcal{T}$ with respect to $\dot{X}_{m,j+1}$ gives

$$\left(\sum_{i=1}^{n} \int_a^b R_i \frac{\partial R_i}{\partial \dot{X}_{m,j+1}} dx \right) + \bar{P}_{m,j} - \epsilon_{m,j} S_{m,j} + \epsilon_{m,j+1} S_{m,j+1} = 0 \tag{2.47}$$

where

$$\begin{aligned} \bar{P}_{m,j} &= -\left(\epsilon_{m,j}\right)^2 \frac{dX_{m,j}}{dt} \\ &\quad + \left[\left(\epsilon_{m,j}\right)^2 + \left(\epsilon_{m,j+1}\right)^2\right] \frac{dX_{m,j+1}}{dt} \\ &\quad - \left(\epsilon_{m,j+1}\right)^2 \frac{dX_{m,j+2}}{dt} \end{aligned} \tag{2.48}$$

Once again, from Equation 2.38 and from Equation 2.24 the second equation of system defined by Equation 2.44 takes the form

$$\sum_{i=1}^{n} \int_a^b R_i g_{i,m} \frac{\partial U_m}{\partial X_{m,j+1}} dx + \bar{P}_{m,j} - \epsilon_{m,j} S_{m,j} + \epsilon_{m,j+1} S_{m,j+1} = 0 \tag{2.49}$$

The terms of Equation 2.49 which involve the internodal viscosity has the effect of preventing relative movements between neighboring nodes and inhibits the mass matrix to be singular. The terms containing the internodal spring functions penalize the absolute distance between neighboring nodes.

Regarding the first equation of system by Equation 2.44, we can highlight two different situations. The first one occurs when the equation is associated with a local interior node. In this case, $U_{m,j}^k$ is the approximate solution at the local interior node $s_{m,j}^k$. The second situation appears when the equation is associated with the approximate solution at a node of the spatial mesh. It is the case where $U_{m,j}^{p_{m,j}} = U_{m,j+1}^1$ is the approximate solution at the node $X_{m,j+1} = s_{m,j}^{p_{m,j}} = s_{m,j+1}^1$. To simplify the notation let us define

$$\mathcal{F}_i = \sum_{l=1}^{n} f_{i,l} \frac{\partial^2 U_l}{\partial x^2} + h_i$$

$$\mathcal{G}_i = \sum_{l=1}^{n} g_{i,l} \frac{\partial U_l}{\partial t} \tag{2.50}$$

The corresponding ODE system has three distinct kinds of equations:

1. **Equation associated with the approximate solution at a local interior node.**

$$\sum_{i=1}^{n} \int_a^b g_{i,m} \frac{\partial U_m}{\partial U_{m,j}^k} \mathcal{G}_i dx = \sum_{i=1}^{n} \int_a^b g_{i,m} \frac{\partial U_m}{\partial U_{m,j}^k} \mathcal{F}_i dx \tag{2.51}$$

2. **Equation associated with the approximate solution at a spatial node.**

$$\sum_{i=1}^{n} \int_a^b g_{i,m} \frac{\partial U_m}{\partial U_{m,j}^{p_{m,j}}} \mathcal{G}_i dx = \sum_{i=1}^{n} \int_a^b g_{i,m} \frac{\partial U_m}{\partial U_{m,j}^{p_{m,j}}} \mathcal{F}_i dx \tag{2.52}$$

3. **Equation associated with a spatial node.**

$$\sum_{i=1}^{n} \int_a^b g_{i,m} \frac{\partial U_m}{\partial X_{m,j+1}} \mathcal{G}_i dx + \bar{P}_{m,j} = \tag{2.53}$$

$$\sum_{i=1}^{n} \int_a^b g_{i,m} \frac{\partial U_m}{\partial X_{m,j+1}} \mathcal{F}_i dx + \epsilon_{m,j} S_{m,j} - \epsilon_{m,j+1} S_{m,j+1}$$

2.2.6 MFEM equations

To express the resulting system of ordinary differential equations the integrals appearing in Equation 2.51, Equation 2.52, and Equation 2.53 must be evaluated. Spatial domain decomposition is used to the calculation of integrals. For each mesh $\mathcal{M}_m$, the spatial domain is set to

$$\Omega = [a, b] = \bigcup_{j=1}^{q_m} \Omega_{m,j} \tag{2.54}$$

where $\Omega_{m,j} = [X_{m,j}, X_{m,j+1}]$.

Since we use different spatial mesh for each component of the solution, we are forced to take into account the spatial nodes of the other spatial meshes belonging to $\Omega_{m,j}$. Consider the ordered set $\Theta_{m,j}$ such that the first and the last element define the *j-th* finite element of the space mesh $\mathcal{M}_m$. The other points of $\Theta_{m,j}$ are nodes from a different mesh $\mathcal{M}_i, i \neq m$ belonging to the interval $[X_{m,j}, X_{m,j+1}]$. This means that if

$$\Theta_{m,j} = \{\Theta_{m,j}^0, \Theta_{m,j}^1, ..., \Theta_{m,j}^{r_{m,j}}, \Theta_{m,j}^{r_{m,j}+1}\} \tag{2.55}$$

then

$$\Theta_{m,j}^0 = X_{m,j} \text{ and } \Theta_{m,j}^{r_{m,j}+1} = X_{m,j+1} \tag{2.56}$$

and the other $r_{m,j}$ points are spatial nodes but do not belong to the mesh $\mathcal{M}_m$. For an illustration of the situation consider a problem with three meshes, $\mathcal{M}_1$ with four finite elements and the others $\mathcal{M}_2$ and $\mathcal{M}_3$ with three finite elements. Figure 2.3 shows the position of spatial nodes at time t. Taking the second finite element of mesh $\mathcal{M}_3$, Figure 2.3 also shows the $\Theta_{3,2}$ set of points to be consider to perform integration on the finite element $\Omega_{3,2}$.

Until now the difficulty in dealing with the spatial derivatives of the approximate solution was ignored. The problem is that the approximate solution U_m to u_m, defined by Equation 2.20, is a piecewise polynomial. Equation 2.17 and Equation 2.18 define the spatial derivatives on the restriction of U_m to the finite element $\Omega_{m,j}$. The spatial derivatives of U_m are not defined at the spatial nodes. To overcome this problem the piecewise polynomial approximation is replaced, in the neighborhood of a node, by a smoothed function having as many continuous spatial derivatives as required by the partial differential equation. Let us consider a sufficiently small neighborhood of a spatial node $X_{m,j}$, and define the smoothed function $\tilde{U}_m$ by taking

$$\tilde{U}_m = \begin{cases} U_m & \text{if } x \notin \mathcal{V}_\delta \\ \mathcal{H} & \text{if } x \in \mathcal{V}_\delta \end{cases} \tag{2.57}$$

where $\mathcal{V}_\delta = [X_{m,j} - \frac{\delta}{2}, X_{m,j} + \frac{\delta}{2}]$ with δ sufficiently small and $\mathcal{H}$ is the cubic

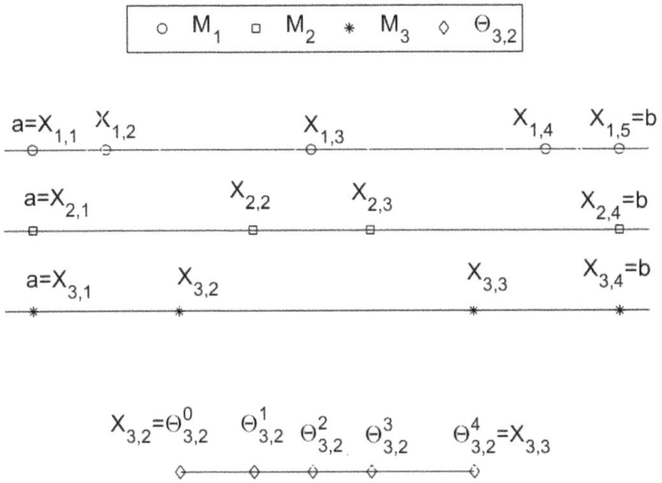

| o M_1 | □ M_2 | * M_3 | ◇ $\Theta_{3,2}$ |

$a=X_{1,1}$ $X_{1,2}$ $X_{1,3}$ $X_{1,4}$ $X_{1,5}=b$

$a=X_{2,1}$ $X_{2,2}$ $X_{2,3}$ $X_{2,4}=b$

$a=X_{3,1}$ $X_{3,2}$ $X_{3,3}$ $X_{3,4}=b$

$X_{3,2}=\Theta_{3,2}^0$ $\Theta_{3,2}^1$ $\Theta_{3,2}^2$ $\Theta_{3,2}^3$ $\Theta_{3,2}^4=X_{3,3}$

FIGURE 2.3
Finite element $\Omega_{3,2}$ partition.

Hermite polynomial satisfying

$$
\begin{aligned}
\mathcal{H}(X_{m,j} - \tfrac{\delta}{2}) &= U_m(X_{m,j} - \tfrac{\delta}{2}) \\
\mathcal{H}(X_{m,j} + \tfrac{\delta}{2}) &= U_m(X_{m,j} + \tfrac{\delta}{2}) \\
\mathcal{H}'(X_{m,j} - \tfrac{\delta}{2}) &= \frac{\partial U_m}{\partial x}(X_{m,j} - \tfrac{\delta}{2}) \\
\mathcal{H}'(X_{m,j} + \tfrac{\delta}{2}) &= \frac{\partial U_m}{\partial x}(X_{m,j} + \tfrac{\delta}{2})
\end{aligned}
\tag{2.58}
$$

as Figure 2.4 illustrates. Briefly, these conditions tell us that when $\delta \to 0$ the smoothed function $\tilde{U}_m$ approximates the piecewise polynomial U_m. Therefore the approximate solution $\tilde{U}_m$ and its spatial derivatives have sufficient continuity properties so that the residual equation 2.38 is well defined.

This approximate solution $\tilde{U}_m$ will be used for the computations concerning the integrals appearing in Equation 2.51, Equation 2.52, and Equation 2.53. For this purpose the finite element $\Omega_{m,j}$ must be decomposed by using the $\Theta_{m,j}$ set of points. Here it is convenient to denote the limiting value $\Theta_{m,j}^r \equiv \Theta_{m,j}^{r^+}$ in the sense that

$$
U_m(\Theta_{m,j}^r) = U_m(\Theta_{m,j}^{r^+}) = U_m(\lim_{\delta \to 0} \Theta_{m,j}^r + \delta/2)
$$

and the limiting value $\Theta_{m,j}^{r+1} \equiv \Theta_{m,j}^{r+1^-}$ is taking in the sense that

$$
U_m(\Theta_{m,j}^{r+1}) = U_m(\Theta_{m,j}^{r+1^-}) = U_m(\lim_{\delta \to 0} \Theta_{m,j}^{r+1} - \delta/2)
$$

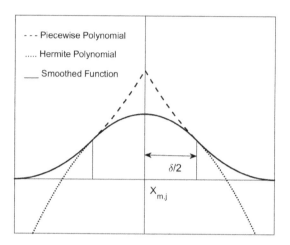

FIGURE 2.4
Smoothed approximate solution in a neighborhood of a spatial node.

The integrals are calculated over each $[\Theta_{m,j}^{r}{}^{+}, \Theta_{m,j}^{r+1}{}^{-}]$ interval and over each V_δ neighborhood of a $\Theta_{m,j}^{r}$ node and taking $\delta \to 0$. The first one is evaluated by a Lobatto quadrature. The exception is when the spatial mesh does not include the boundary nodes $X_{m,1}$ or X_{m,q_m+1}. This is a feature of our code. The user may choose to include or not include the endpoints of the spatial domain on the spatial mesh. In such cases a Radau quadrature formula, from the left or from the right, is considered. For these integrals the truncation error is null by using an appropriate number of quadrature points. Over each V_δ it is sufficient to take a quadrature rule such as the midpoint rule. Before proceeding we must clarify the notation used. Let $\Theta_{m,j}^{r}$ be a point of the set $\Theta_{m,j}$. Then $\Theta_{m,j}^{r} = X_{l,J+1}$ for some mesh $\mathcal{M}_l$. We will designate by

$$\left. \frac{\partial U_l}{\partial x} \right|_{\Theta_{m,j}^{r}{}^{+}} \tag{2.59}$$

the spatial derivative of the restriction U_l to the finite element $\Omega_{l,J+1}$ calculated at $X_{l,J+1}$ and

$$\left. \frac{\partial U_l}{\partial x} \right|_{\Theta_{m,j}^{r}{}^{-}} \tag{2.60}$$

the spatial derivative of the restriction U_l to the finite element $\Omega_{l,J}$ calculated at $X_{l,J+1}$.

Two integrals appear in Equation 2.51, Equation 2.52, and Equation 2.53 that have a major relevance. The first is

$$F_{m,j}^k = \frac{1}{Z_{m,j}} \sum_{i=1}^{n} \int_{X_{m,j}^+}^{X_{m,j+1}^-} g_{i,m}\, \ell_{m,j}^k(x) \mathcal{F}_i dx \tag{2.61}$$

The second is the integral defined on a neighborhood of a spatial node $X_{m,j+1}$

$$E_{m,j+1} = \int_{X_{m,j+1}-\delta/2}^{X_{m,j+1}+\delta/2} g_{i,m} \frac{\partial U_m}{\partial U_{m,j}^{p_{m,j}}} \mathcal{F}_i\, dx \tag{2.62}$$

The integrals on a V_δ neighborhood of a spatial node, for which the integrand function does not include the second spatial derivatives, are zero in the limit when $\delta \to 0$. This result is obtained by using a quadrature rule that transforms the integral into a product of a non-null factor by δ. For those involving the second spatial derivatives, a quadrature rule must be used. It is sufficient to take a quadrature rule such as the midpoint rule and then take the limit $\delta \to 0$. This leads to

$$Z_{m,j} F_{m,j}^k = \sum_{r=0}^{r_{m,j}} \sum_{i=1}^{n} \int_{\Theta_{m,j}^r{}^+}^{\Theta_{m,j}^{r+1}{}^-} g_{i,m}\, \ell_{m,j}^k(x)\, \mathcal{F}_i\, dx \tag{2.63}$$

$$+ \sum_{r=1}^{r_{m,j}} \sum_{l=1}^{n} \left(\frac{\partial U_l}{\partial x} \Big|_{\Theta_{m,j}^r{}^+} - \frac{\partial U_l}{\partial x} \Big|_{\Theta_{m,j}^r{}^-} \right) \sum_{i=1}^{n} \left[\ell_{m,j}^k(x)\, g_{i,m}\, f_{i,l} \right] \Big|_{\Theta_{m,j}^r}$$

and

$$E_{m,j+1} = \sum_{l=1}^{n} \left(\frac{\partial U_l}{\partial x} \Big|_{X_{m,j+1}^+} - \frac{\partial U_l}{\partial x} \Big|_{X_{m,j+1}^-} \right) \sum_{i=1}^{n} (g_{i,m}\, f_{i,l}) \Big|_{X_{m,j+1}} \tag{2.64}$$

Notice that the value of the expression

$$\frac{\partial U_l}{\partial x} \Big|_{\Theta_{m,j}^r{}^+} - \frac{\partial U_l}{\partial x} \Big|_{\Theta_{m,j}^r{}^-}$$

appearing in Equation 2.63 depends on the position of the spatial node $\Theta_{m,j}^r$ on the mesh $\mathcal{M}_l$. Assuming that $\Theta_{m,j}^r$ is not on the boundary of the spatial domain, then two cases can occur:

- If $\Theta_{m,j}^r \notin \mathcal{M}_l$ then the expression vanishes.

- If $\Theta_{m,j}^r \in \mathcal{M}_l$, i.e. $\Theta_{m,j}^r = X_{l,J+1}$ then

$$\frac{\partial U_l}{\partial x} \Big|_{\Theta_{m,j}^r{}^+} - \frac{\partial U_l}{\partial x} \Big|_{\Theta_{m,j}^r{}^-} = \frac{\partial U_{l,J+1}}{\partial x} \Big|_{s_{l,J+1}^1} - \frac{\partial U_{l,J}}{\partial x} \Big|_{s_{l,J}^{p_{l,J}}} \tag{2.65}$$

We are now in position to define the set of MFEM equations.

1. **Equation associated with the approximate solution at a local interior node.** In this case $U_{m,j}^k$ is the approximate solution at the local interior node $s_{m,j}^k$. By Equation 2.27 and Equation 2.28 all integrals are null except for the finite element $\Omega_{m,j}$ so Equation 2.51 becomes,

$$\sum_{r=0}^{r_{m,j}} \sum_{i=1}^{n} \int_{\Theta_{m,j}^{r+}}^{\Theta_{m,j}^{r+1-}} g_{i,m} \; \ell_{m,j}^k(x) \; \mathcal{G}_i dx = Z_{m,j} F_{m,j}^k \qquad (2.66)$$

2. **Equation associated with the approximate solution at a spatial node.** For the moment, let us suppose that the node is not on the boundary of spatial domain. In this case $U_{m,j}^{p_{m,j}} = U_{m,j+1}^1$ is the approximate solution at the node $X_{m,j+1} = s_{m,j}^{p_{m,j}} = s_{m,j+1}^1$. By Equation 2.27 and Equation 2.29, all integrals are null except for $\Omega_{m,j} \cup \Omega_{m,j+1}$ and Equation 2.52 gives

$$\sum_{r=0}^{r_{m,j}} \sum_{i=1}^{n} \int_{\Theta_{m,j}^{r+}}^{\Theta_{m,j}^{r+1-}} g_{i,m} \; \ell_{m,j}^{p_{m,j}}(x) \; \mathcal{G}_i \; dx$$

$$+ \sum_{r=0}^{r_{m,j+1}} \sum_{i=1}^{n} \int_{\Theta_{m,j+1}^{r+}}^{\Theta_{m,j+1}^{r+1-}} g_{i,m} \; \ell_{m,j+1}^1(x) \; \mathcal{G}_i \; dx \qquad (2.67)$$

$$= Z_{m,j} F_{m,j}^{p_{m,j}} + Z_{m,j+1} F_{m,j+1}^1 + E_{m,j+1}$$

3. **Equation associated with a spatial node.** Since the physical spatial domain is fixed, consider $X_{m,j+1}$, $j = 1, \cdots, q_m - 1$. From Equation 2.17, Equation 2.36, and Equation 2.48 the Equation 2.53 can be written as

$$H_{m,j} \sum_{r=0}^{r_{m,j}} \sum_{i=1}^{n} \int_{\Theta_{m,j}^{r+}}^{\Theta_{m,j}^{r+1-}} g_{i,m} \; \ell_{m,j}^{p_{m,j}}(x) \; \mathcal{G}_i dx$$

$$-H_{m,j} \sum_{r=0}^{r_{m,j+1}} \sum_{i=1}^{n} \int_{\Theta_{m,j+1}^{r+}}^{\Theta_{m,j+1}^{r+1-}} g_{i,m} \; \ell_{m,j+1}^1(x) \; \mathcal{G}_i dx$$

$$= H_{m,j} \left(Z_{m,j} F_{m,j}^{p_{m,j}} - Z_{m,j+1} F_{m,j+1}^1 \right) \qquad (2.68)$$

$$+\epsilon_{m,j} S_{m,j} - \epsilon_{m,j+1} S_{m,j+1} - \bar{P}_{m,j}$$

where

$$H_{m,j} = \frac{1}{2}\left(\left.\frac{\partial U_{m,j+1}}{\partial x}\right|_{s^1_{m,j+1}} - \left.\frac{\partial U_{m,j}}{\partial x}\right|_{s^{p_{m,j}}_{m,j}}\right). \tag{2.69}$$

2.2.7 Boundary conditions

Until now we have ignored the boundary conditions associated with the model Equation 2.6. After having understood the structure of the systems of ordinary equations produced by the MFEM the modifications introduced by the boundary conditions are quite simple. Let us consider the general formula for the boundary condition

$$\gamma^a_m \left.\frac{\partial u_m}{\partial x}\right|_{x=a} = \mu^a_m u_m(a,t) + \nu^a_m, \qquad t \geq 0$$

$$\gamma^b_m \left.\frac{\partial u_m}{\partial x}\right|_{x=b} = \mu^b_m u_m(b,t) + \nu^b_m, \qquad t \geq 0 \tag{2.70}$$

which includes Dirichlet, Neumann, or Robin boundary conditions. This class of boundary conditions covers a large number of models that frequently arise in the study of transport problems in porous media, such as separation and reaction processes.

Dirichlet boundary conditions. For models with Dirichlet boundary conditions the solution is known and constant at the boundary of the physical domain. Thus in Equation 2.70, $\gamma^a_m = 0$ tells us that a Dirichlet boundary condition is assumed at the left end of the domain. Analogously, a Dirichlet boundary condition at the right end of the domain is imposed by setting $\gamma^b_m = 0$. As a consequence of this condition, we can state the following changes on MFEM equations:

- In Equation 2.24 that defines the temporal derivative of the approximate solution U_m we will have

$$\gamma^a_m = 0 \Leftrightarrow \frac{dU^1_{m,1}}{dt} = 0 \tag{2.71}$$

$$\gamma^b_m = 0 \Leftrightarrow \frac{dU^{p_m,q_m}_{m,q_m}}{dt} = 0$$

So, the last two terms of 2.24 will be null whenever a Dirichlet condition is associated with one of the ends.

- In the Equation 2.42 if a Dirichlet condition is associated with one or both ends of spatial domain the corresponding time derivative is not an unknown to be determined.

- There are no equations associated with the approximate solution at a spatial node, Equation 2.52 and Equation 2.67, if the spatial node is on the boundary of the physical domain and a Dirichlet condition is associated with it.

Time-dependent Dirichlet boundary conditions. If time-dependent Dirichlet boundary conditions are given then the time derivative at the boundary points is defined. Then in Equation 2.24, Equation 2.52, and Equation 2.67 the corresponding value of the time derivative is used.

Robin boundary conditions. Neumann and Robin boundary conditions are treated together, assuming that $\gamma_m^a = 1$ or $\gamma_m^b = 1$. The major implications of this kind of boundary is in Equation 2.67. If it is the case of $\gamma_m^a = 1$ then Equation 2.67, associated with $U_{m,1}^1$, is of the form

$$\sum_{r=0}^{r_{m,1}} \sum_{i=1}^{n} \int_{\Theta_{m,1}^r{}^+}^{\Theta_{m,1}^{r+1}{}^-} g_{i,m}\, \ell_{m,1}^1(x)\, \mathcal{G}_i dx = Z_{m,1} F_{m,1}^1 + E_{m,1} \qquad (2.72)$$

For the evaluation of $E_{m,1}$ we note that, if $\gamma_l^a = 1$ then

$$\left.\frac{\partial U_l}{\partial x}\right|_{X_{m,1}^+} - \left.\frac{\partial U_l}{\partial x}\right|_{X_{m,1}^-} = \left.\frac{\partial U_{l,1}}{\partial x}\right|_{s_{l,1}^1} -\mu_l^a U_{l,1}^1 - \nu_l^a \qquad (2.73)$$

and otherwise, $\left.\frac{\partial U_l}{\partial x}\right|_{X_{m,1}^+} - \left.\frac{\partial U_l}{\partial x}\right|_{X_{m,1}^-} = 0$. This also applies to the calculation of term $F_{m,1}^1$.

For the same reason, if it is the case of $\gamma_m^b = 1$ then Equation 2.67, associated with $U_{m,q_m}^{p_{m,q_m}}$, is of the form

$$\sum_{r=0}^{r_{m,q_m}} \sum_{i=1}^{n} \int_{\Theta_{m,q_m}^r{}^+}^{\Theta_{m,q_m}^{r+1}{}^-} g_{i,m}\, \ell_{m,q_m}^{p_{m,q_m}}(x)\mathcal{G}_i dx = Z_{m,q_m} F_{m,q_m}^{p_{m,q_m}} + E_{m,q_m+1} \qquad (2.74)$$

Concerning the evaluation of E_{m,q_m+1} and $F_{m,q_m}^{p_{m,q_m}}$ terms, if $\gamma_l^b = 1$ then

$$\left.\frac{\partial U_l}{\partial x}\right|_{X_{m,q_m+1}^+} - \left.\frac{\partial U_l}{\partial x}\right|_{X_{m,q_m+1}^-} = \mu_l^b U_{l,q_l}^{p_{l,q_l}} + \nu_l^b - \left.\frac{\partial U_{l,q_l}}{\partial x}\right|_{s_{l,q_l}^{p_{l,q_l}}} \qquad (2.75)$$

and otherwise this expression vanishes.

No boundary conditions. If there are no boundary conditions, then in Equation 2.72 and Equation 2.74 the terms $E_{m,1}$ or E_{m,q_m+1} or both will vanish.

2.2.8   Mass matrix structure

The semi-discretization of the partial differential system 2.6 by the MFEM produces a system of ordinary differential equation to be solved. Here we make an analysis of the mass matrix to understand the characteristics of this matrix.

Consider a system of two partial differential equations with two dependent variables subject to Dirichlet boundary condition. Assume that the endpoints of the spatial domain are spatial nodes of the meshes $\mathcal{M}_1$ and $\mathcal{M}_2$. The generation of the computational meshes is performed according the conditions expressed in Table 2.1. The user may choose the number of finite elements and the number of local interior nodes in each element.

TABLE 2.1
Conditions used for mesh generation in Figure 2.5.

Meshes Properties	$\mathcal{M}_1$	$\mathcal{M}_2$
Number of finite elements	3	4
Number of local interior nodes	5; 2; 5	3; 2; 2; 3

The spatial discretization with the conditions expressed in Table 2.1 produces a system of 36 ordinary differential equations. Figure 2.5 shows the spatial nodes for the two meshes as well as the position of the local interior nodes at a time t. The enumeration of spatial nodes used to define each equation of the ODE system is also indicated in Figure 2.5.

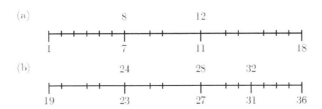

FIGURE 2.5
Spatial meshes and local interior nodes (a) $\mathcal{M}_1$ (b) $\mathcal{M}_2$.

The structure of the mass matrix is shown in Figure 2.6. The nonzero blocks along the diagonal maintain the initial structure because the number of nodes is constant in time. However the same is not true for blocks outside the diagonal whose positions vary with time.

2.2.9   MFEM equations for explicit systems

We observe that some problems are modeled by an explicit system of partial differential equations. This corresponds to having the identity matrix in the place of $\mathbf{G}$ in Equation 2.6. For explicit problems, we can obtain a simpler formulation for the general MFEM Equation 2.51, Equation 2.52, and

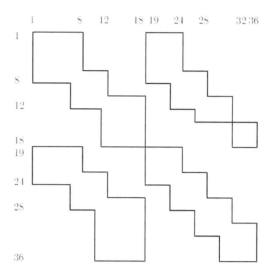

FIGURE 2.6
Mass matrix: Nonzero blocks at time t.

Equation 2.53. For explicit systems the two integrals 2.63 and 2.64 are

$$Z_{m,j} F_{m,j}^k = \sum_{r=0}^{r_{m,j}} \int_{\Theta_{m,j}^{r}{}^+}^{\Theta_{m,j}^{r+1}{}^-} \ell_{m,j}^k(x)\, \mathcal{F}_m \, dx \tag{2.76}$$

$$+ \sum_{r=1}^{r_{m,j}} \sum_{l=1}^{n} \left(\frac{\partial U_l}{\partial x}\Big|_{\Theta_{m,j}^r{}^+} - \frac{\partial U_l}{\partial x}\Big|_{\Theta_{m,j}^r{}^-} \right) \left[\ell_{m,j}^k(x)\, f_{m,l} \right]\Big|_{\Theta_{m,j}^r}$$

and

$$E_{m,j+1} = \sum_{l=1}^{n} \left(\frac{\partial U_l}{\partial x}\Big|_{X_{m,j+1}^+} - \frac{\partial U_l}{\partial x}\Big|_{X_{m,j+1}^-} \right) f_{m,l}\Big|_{X_{m,j+1}} \tag{2.77}$$

By using the reference element, let us define

$$I_{m,j}^{k,i} = \int_{0+}^{1^-} \ell_{m,j}^k(\xi_{m,j})\ell_{m,j}^i(\xi_{m,j})\, d\xi, \tag{2.78}$$

$$C_{m,j}^k = \sum_{i=1}^{q_{m,j}} I_{m,j}^{k,i} \xi_{m,j}^i \frac{\partial U_{m,j}}{\partial x}\Big|_{s_{m,j}^i} \tag{2.79}$$

and

$$D_{m,j}^k = \sum_{i=1}^{q_{m,j}} I_{m,j}^{k,i} \left(1 - \xi_{m,j}^i\right) \frac{\partial U_{m,j}}{\partial x}\Big|_{s_{m,j}^i} \tag{2.80}$$

According to Sereno [41], taking the quantities $I_{m,j}^{k,i}$, $C_{m,j}^k$ and $D_{m,j}^k$ the space discretization produces a system of ordinary differential equations as indicated below:

- **Equation associated with the approximate solution at a local interior node.**

  In this case $U_{m,j}^k$ is the approximate solution at local node $s_{m,j}^k$. Equation 2.66 becomes

  $$\sum_{i=1}^{p_{m,j}} I_{m,j}^{i,k} \frac{dU_{m,j}^i}{dt} - D_{m,j}^k \frac{dX_{m,j}}{dt} - C_{m,j}^k \frac{dX_{m,j+1}}{dt} = F_{m,j}^k \qquad (2.81)$$

- **Equation associated with the approximate solution at a spatial node.**

  In this case $U_{m,j}^{p_{m,j}} = U_{m,j+1}^1$ is the approximate solution at the node $X_{m,j+1} = s_{m,j}^{p_{m,j}} = s_{m,j+1}^1$. Assuming that the node is not on the boundary, Equation 2.67 gives

  $$Z_{m,j}\left(\sum_{i=1}^{q_{m,j}} I_{m,j}^{q_{m,j},i} \frac{dU_{m,j}^i}{dt} - D_{m,j}^{q_{m,j}} \frac{dX_{m,j}}{dt} - C_{m,j}^{q_{m,j}} \frac{dX_{m,j+1}}{dt}\right)$$

  $$+ Z_{m,j+1} \sum_{i=1}^{q_{m,j+1}} I_{m,j+1}^{1,i} \frac{dU_{m,j+1}^i}{dt}$$

  $$+ Z_{m,j+1}\left(-D_{m,j+1}^1 \frac{dX_{m,j+1}}{dt} - C_{m,j+1}^1 \frac{dX_{m,j+1}}{dt}\right)$$

  $$= Z_{m,j} F_{m,j}^{p_{m,j}} + Z_{m,j+1} F_{m,j+1}^1 + E_{m,j+1} \qquad (2.82)$$

- **Equation associated with a spatial node.**

  Let $X_{m,j+1}$, $j = 1, \cdots, q_m - 1$ be an interior spatial node. Equation 2.68 becomes

  $$H_{m,j} Z_{m,j} \sum_{i=1}^{q_{m,j}} I_{m,j}^{q_{m,j},i} \frac{dU_{m,j}^i}{dt}$$

  $$H_{m,j} Z_{m,j}\left(-D_{m,j}^{q_{m,j}} \frac{dX_{m,j}}{dt} - C_{m,j}^{q_{m,j}} \frac{dX_{m,j+1}}{dt}\right)$$

  $$- H_{m,j} Z_{m,j+1} \sum_{i=1}^{q_{m,j+1}} I_{m,j+1}^{1,i} \frac{dU_{m,j+1}^i}{dt}$$

  $$- H_{m,j} Z_{m,j+1}\left(-D_{m,j+1}^1 \frac{dX_{m,j+1}}{dt} - C_{m,j+1}^1 \frac{dX_{m,j+1}}{dt}\right)$$

  $$= H_{m,j}\left(Z_{m,j} F_{m,j}^{p_{m,j}} - Z_{m,j+1} F_{m,j+1}^1\right)$$

  $$+ \epsilon_{m,j} S_{m,j} - \epsilon_{m,j+1} S_{m,j+1} - \bar{P}_{m,j} \qquad (2.83)$$

As mentioned in Section 2.2.7 the boundary conditions imposed can suppress or change these equations according to their nature.

2.2.10   Time integration

To obtain the fully discrete numerical solutions and the meshes where the solutions are represented, it is necessary to solve the resulting system of ordinary equations. Usually the spatial discretization produces a stiff system of ODEs. Here we take advantage of the availability of high-quality ODE solvers. In fact, a vast collection of quality ODE solvers is available. Examples include DASSL [48], the suite of programs for solving ODE in MATLAB® [49] and Radau methods [50]. For that purpose, in the 1D-MFEM numerical code, we used the package LSODIS developed at the Lawrence Livermore National Laboratory [51]. LSODIS is included on ODEPACK [52], a collection of FORTRAN solvers for initial value problems for ordinary differential equations.

2.3   The MFEM in Higher Dimensions

Guided by the successful results in solving time-dependent problems in one-dimensional space with the moving finite elements method, it is now tempting to extend the method to problems in higher-dimensional spatial domains. In this section, we shall concentrate our attention on time-dependent partial differential problems where the spatial variable lies in a space of dimension two. Mathematical models for chemical engineering problems, such as adsorption in a fixed-bed, give us some clues on how to extend the moving finite element. They indicate that there are essentially two possibilities to extend the idea presented in Section 2.2, which is so successful, into the two-dimensional space.

The first one is the obvious one. Instead of discretizing a 1D spatial domain, we shall discretize a 2D spatial domain. In Section 2.3.1, we introduce the modifications needed to the 1D MFEM so that it can be used to solve time-dependent partial differential problems in two-dimensional space domains. We shall detail questions related to the derivative of the approximate solution with respect to the mesh deformation. Also the smoothing procedure to ensure sufficient conditions of differentiability is discussed in detail.

The second one takes advantage of the structure of the problem and applies to problems in a two-dimensional spatial domain with a specific 1D+1d structure, wherever it is possible to identify a macroscale and a microscale. This new numerical procedure takes advantage of the well-known abilities of the 1D-MFEM and looks at the particular structure of the problem to produce accurate solutions with less computational effort. Section 2.3.2 presents the MFEM strategy to solve 1D+1d time-dependent partial differential equations.

2.3.1 MFEM formulation in two-dimensional (2D) space domains

In this section, we consider the formulation of the moving finite elements method in a two-dimensional space domain. The ideas and structure of MFEM for 2D models are similar to those developed for models in 1D spatial domains. However, the implementation of those ideas is more complex. For convenience, let us consider a rectangular spatial domain, $\Omega = [a, b] \times [c, d]$. We shall denote the physical boundary of Ω by $\partial\Omega$ and the set of all interior points by $\overset{o}{\Omega}$.

Our formulation of MFEM still holds with only few changes to the case of irregular domains. In Chapter 6 we will give an alternative development of the MFEM for time-dependent partial differential equations on 2D irregular spatial domains [45]. Furthermore, the use of different meshes associated with each of the dependent variables, in the case of systems of partial differential equations, will be considered later.

Let consider an explicit two-dimensional convection–diffusion–reaction system with general *m-th* equation has the following form

$$\frac{\partial u_m}{\partial t} = \sum_{k_e=1}^{n} \left[f_{m,k_e}\left(X, t, \mathbf{u}, \frac{\partial \mathbf{u}}{\partial x}, \frac{\partial \mathbf{u}}{\partial y}\right) \frac{\partial^2 u_{k_e}}{\partial x^2}\right] + \tag{2.84}$$

$$+ \sum_{k_e=1}^{n} \left[g_{m,k_e}\left(X, t, \mathbf{u}, \frac{\partial \mathbf{u}}{\partial x}, \frac{\partial \mathbf{u}}{\partial y}\right) \frac{\partial^2 u_{k_e}}{\partial y^2}\right] + h_m\left(X, t, \mathbf{u}, \frac{\partial \mathbf{u}}{\partial x}, \frac{\partial \mathbf{u}}{\partial y}\right)$$

where $t \geq 0$ and $X = (x, y) \in \Omega$, subject to Dirichlet, Neumann, or Robin boundary conditions and appropriated initial condition,

$$u_m(X, 0) = u_{0_m}(X)$$

and $\mathbf{u}(X, t) = [u_1(X, t), u_2(X, t), \cdots, u_n(X, t)]^T$ is the vector of dependent variables.

Spatial discretization

The spatial discretization of domain Ω uses a standard finite element triangulation

$$\Omega = \bigcup_{j=1}^{q} \Omega_j \tag{2.85}$$

so that each spatial node is the vertex of at least one triangle Ω_j, and the intersection of two triangles is either:

- the empty set (the two triangles are distinct)

- a vertex (the two triangles share a vertex)

- an edge (the two triangles share an edge).

Spatial nodes at the border, $\partial\Omega$, are vertices of three triangles, except for corners that are vertices of only one or two triangles. In addition, spatial nodes not belonging to $\partial\Omega$ are the vertex of six triangles. In this formulation we assume that the spatial mesh $\mathcal{M}$ is the same for all dependent variables. Let n_s be the number of spatial nodes of $\mathcal{M}$.

$$\mathcal{M} = \{X_i(t) \in \Omega : X_i = (x_i(t), y_i(t)), i = 1, ..., n_s\} \tag{2.86}$$

where

$$a = x_1 < x_2 < ... < x_{n_x+1} = b \tag{2.87}$$

and

$$c = y_1 < y_2 < ... < y_{n_y+1} = d \tag{2.88}$$

The number of spatial nodes is $n_s = (n_x + 1)(n_y + 1)$ and the number of finite elements is $q = 2n_x n_y$.

The spatial mesh is time-dependent. However we dropped the time label for simplicity of the notation. In addition, there is no need to use the subscript identifying the dependent variable. This is valid for the mesh $\mathcal{M}$, for the number of triangles, q, as well as for other variables that we will use later. An example of an initial triangulation of the spatial domain $\Omega = [0, 1] \times [0, 1]$ with 60 finite elements can be visualized in Figure 2.7.

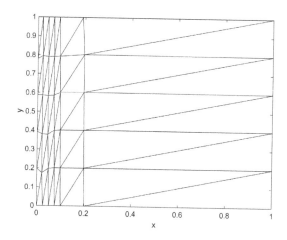

FIGURE 2.7
Initial mesh on $\Omega = [0, 1] \times [0, 1]$ with 60 finite elements.

Element-by-element approach

In an analogous manner to the one-dimensional case, we will write the approximate solution in a finite element, Ω_j. To do this, we will consider a reference element and redundant natural coordinates (L_1, L_2, L_3) with $L_1 +$

$L_2 + L_3 = 1$. For a given finite element Ω_j with Cartesian coordinates X_{j_1}, X_{j_2}, and X_{j_3}, the natural coordinates of the vertices are, respectively, $(1, 0, 0)$, $(0, 1, 0)$, and $(0, 0, 1)$. The relation between the Cartesian coordinates of a point $X = (x, y) \in \Omega_j$ and its natural coordinates L_1, L_2, L_3 is defined by

$$x = x_{j_1}L_1 + x_{j_2}L_2 + x_{j_3}L_3 \qquad (2.89)$$

and

$$y = y_{j_1}L_1 + y_{j_2}L_2 + y_{j_3}L_3 \qquad (2.90)$$

In a matrix notation, we may write,

$$\begin{bmatrix} x \\ y \\ 1 \end{bmatrix} = \begin{bmatrix} x_{j_1} & x_{j_2} & x_{j_3} \\ y_{j_1} & y_{j_2} & y_{j_3} \\ 1 & 1 & 1 \end{bmatrix} \begin{bmatrix} L_1 \\ L_2 \\ L_3 \end{bmatrix} \qquad (2.91)$$

If node numbers are assigned counterclockwise as Figure 2.8 shows then the value of the determinant of matrix $\mathcal{D}$ is twice the area of the triangle, $A(\Omega_j)$,

$$|\mathcal{D}| = \det(\mathcal{D}) = \det \begin{bmatrix} x_{j_1} & x_{j_2} & x_{j_3} \\ y_{j_1} & y_{j_2} & y_{j_3} \\ 1 & 1 & 1 \end{bmatrix} = 2A(\Omega_j) \neq 0 \qquad (2.92)$$

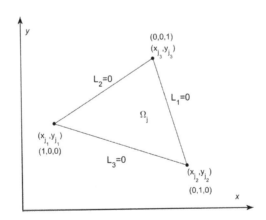

FIGURE 2.8
Natural coordinates and counterclockwise numbering of nodes in Ω_j.

If $X \in \Omega_j$ with Cartesian coordinates (x, y) then the natural coordinates are

$$L_N = \frac{1}{|\mathcal{D}|}(a_N + \Delta y_N x + \Delta x_N y) \quad N = 1, 2, 3 \qquad (2.93)$$

where

$$\begin{cases} a_1 = x_{j_2} y_{j_3} - x_{j_3} y_{j_2} \\ a_2 = x_{j_3} y_{j_1} - x_{j_1} y_{j_3} \\ a_3 = x_{j_1} y_{j_2} - x_{j_2} y_{j_1} \end{cases} \qquad (2.94)$$

and

$$\begin{cases} \Delta y_1 = y_{j_2} - y_{j_3} \\ \Delta y_2 = y_{j_3} - y_{j_1} \\ \Delta y_3 = y_{j_1} - y_{j_2} \end{cases} \begin{cases} \Delta x_1 = x_{j_3} - x_{j_2} \\ \Delta x_2 = x_{j_1} - x_{j_3} \\ \Delta x_3 = x_{j_2} - x_{j_1} \end{cases} \qquad (2.95)$$

Notice that the value of the determinant is

$$|\mathcal{D}| = a_1 + a_2 + a_3 \qquad (2.96)$$

and

$$\Delta x_1 + \Delta x_2 + \Delta x_3 = \Delta y_1 + \Delta y_2 + \Delta y_3 = 0 \qquad (2.97)$$

As described in Section 2.2, we seek a restriction of the approximate solution U_m to u_m in a finite element Ω_j. The construction of the piecewise polynomial of degree n_d is accomplished with the help of Lagrange polynomials. First we have to choose the $p = (n_d + 2)(n_d + 1)/2$ interpolation points in Ω_j and then, for $X \in \Omega_j$ we may define the restriction $U_m(X, t), X \in \Omega_j$ as

$$U_m(X, t) = U_{m,j}(X, t) = \sum_{k=1}^{p} \ell_j^k(X) U_{m,j}^k(t) \qquad (2.98)$$

where $U_{m,j}^k$ is the value of the approximate solution evaluated in the *k-th* interpolation point of *j-th* finite element. The basis functions [53] are given by

$$\ell_j^k = B^{(k)}(L_1) B^{(k)}(L_2) B^{(k)}(L_3) \qquad (2.99)$$

where

$$B^{(k)}(L_N) = \begin{cases} \prod_{l=1}^{d} \frac{1}{l}(n_d L_N - l + 1) & \text{if} \quad d \geq 1 \\ 1 & \text{if} \quad d = 0 \end{cases} \qquad (2.100)$$

with $d = n_d L_N^{(k)}$, $N = 1, 2, 3$ and $k = 1, ..., p$. $L_N^{(k)}$ represents the value of the *N-th* natural coordinate at the *k-th* interpolation point, s_j^k, on finite element Ω_j.

A simple procedure is used to define the interpolation points or local spatial nodes. For linear approximation the interpolation points are the three vertices of the triangle. For a second-degree polynomial on element Ω_j we must have $n_d = 2$ and $p = 6$. Local spatial nodes are placed at the vertices and at the midpoints of each edge of Ω_j. The nodes numbering scheme is counterclockwise starting with the nodes placed at the vertices and then switches to those on the edges. Choosing a polynomial of degree n_d, there are $n_d - 1$ nodes placed on each edge, excluding the vertices. For example, if $n_d = 3$ then $p = 10$. The vertex nodes are numbered 1, 2, and 3. There are six points on the edges, two in each edge and they are numbered from 4 to 9. The centroid is numbered as node 10.

The spatial derivatives of $U_{m,j}$ are also polynomials. Therefore they can be expressed using the local nodes s_j^k on Ω_j. For $X \in \Omega_j$ the spatial derivatives are

$$\frac{\partial U_{m,j}}{\partial x} = \sum_{k=1}^{p} \frac{\partial \ell_j^k}{\partial x}(X) U_{m,j}^k \qquad (2.101)$$

and

$$\frac{\partial U_{m,j}}{\partial y} = \sum_{k=1}^{p} \frac{\partial \ell_j^k}{\partial y}(X) U_{m,j}^k \qquad (2.102)$$

Combining Equation 2.99, Equation 2.100, and Equation 2.101 and by chain rule, it is possible to write,

$$\frac{\partial \ell_j^k}{\partial x} = \sum_{N=1}^{3} \frac{\partial L_N}{\partial x} \frac{\partial}{\partial L_N} \left(B^{(k)}(L_N) \right) \left(\prod_{\substack{K=1 \\ K \neq N}}^{3} B^{(k)}(L_K) \right) \qquad (2.103)$$

and from Equation 2.99, Equation 2.100, and Equation 2.102

$$\frac{\partial \ell_j^k}{\partial y} = \sum_{N=1}^{3} \frac{\partial L_N}{\partial y} \frac{\partial}{\partial L_N} \left(B^{(k)}(L_N) \right) \left(\prod_{\substack{K=1 \\ K \neq N}}^{3} B^{(k)}(L_K) \right) \qquad (2.104)$$

Regarding the second derivatives, we can write, in a condensed form,

$$\frac{\partial^2 U_{m,j}}{\partial z \partial \rho} = \sum_{k=1}^{p} \frac{\partial^2 \ell_j^k}{\partial z \partial \rho}(X) U_{m,j}^k \qquad (2.105)$$

where z is either x or y and the same for ρ.

Global approach

Based on the process described for the one-dimensional space, the global approximate solution is obtained from the approximate solution defined in each element. As before we denote by Δ the set of all local nodes of all finite elements,

$$\Delta = \bigcup_{j=1}^{q} \{s_j^k, k = 1, ..., p\} \qquad (2.106)$$

and since only one mesh is used there is no need to use the subscript identifying the dependent variable.

The global approximation U_m to u_m on Ω is

$$U_m(X,t) = \sum_{i=1}^{D} \Phi_i(X) U_m(\Delta_i) \qquad (2.107)$$

where Φ_i is the global basis function verifying Equation 2.21 and D is the number of global nodes.

This global approximate solution depends on the spatial position of the nodes,

$$X_j = (x_j, y_j), \; j = 1, \cdots, n_s \qquad (2.108)$$

and also depends on the value of the solution at each local node in each of the finite elements,

$$U_{m,j}^k, \; j = 1, \cdots, q, \; k = 1, \cdots, p \qquad (2.109)$$

Moreover, it is convenient to define the support of a global node Δ_i by considering all finite elements to which the point belongs. We shall denote by Ω^i the support of global node Δ_i,

$$\Omega^i = \bigcup_{l=1}^{I} \Omega_l \qquad (2.110)$$

where Ω_l is a finite element verifying $\Delta_i \in \Omega_l$ as Figure 2.9 illustrates.

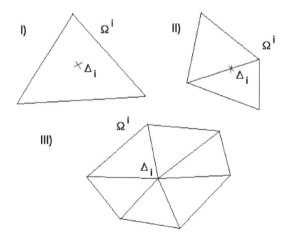

FIGURE 2.9
Support of a global node Δ_i. Distinct cases.

Since X_l and $U_{m,j}^k$ are both time dependent, by the chain rule, the time derivative of U_m will be of the form

$$\frac{\partial U_m}{\partial t} = \sum_{j=1}^{q} \sum_{k=1}^{p} \frac{\partial U_m}{\partial U_{m,j}^k} \frac{dU_{m,j}^k}{dt} + \sum_{l=1}^{n_s} \frac{\partial U_m}{\partial x_l} \frac{dx_l}{dt} + \sum_{l=1}^{n_s} \frac{\partial U_m}{\partial y_l} \frac{dy_l}{dt} \qquad (2.111)$$

In complete analogy to the one-dimensional case, the derivatives with respect to the mesh deformation must be evaluated.

Firstly, consider the derivative $\frac{\partial U_m}{\partial U^k_{m,j}}$. Let $s^k_j \in \Omega_j$ be the local node associated to Δ_i. Thus, $U_m(\Delta_i) = U^k_{m,j}$ and from 2.107

$$\frac{\partial U_m}{\partial U^k_{m,j}} = \Phi_i \qquad (2.112)$$

Besides, we can state that

1. If $\Delta_i \in \overset{o}{\Omega}_j$ then

$$\Phi_i(X) = \begin{cases} 0 & \text{if } X \notin \Omega_j \\ \ell^k_j & \text{if } X \in \Omega_j \end{cases} \qquad (2.113)$$

   where ℓ^k_j is defined by Equation 2.99.

2. If $\Delta_i \in \partial\Omega_j$ and Δ_i is not a spatial node, then Δ_i belongs to the common edge of two triangles, namely, $\Omega_{l_1} \bigcup \Omega_{l_2}$. So,

$$\Phi_i(X) = \begin{cases} 0, & \text{if } X \notin (\Omega_{l_1} \bigcup \Omega_{l_2}) \\ \ell^{k_1}_{l_1}, & \text{if } X \in \Omega_{l_1} \\ \ell^{k_2}_{l_2} & \text{if } X \in \Omega_{l_2} \end{cases} \qquad (2.114)$$

3. If $\Delta_i \in \partial\Omega_j$ and Δ_i is a spatial node, then

$$\Phi_i(X) = \begin{cases} 0, & \text{if } X \notin \overset{I}{\underset{s=1}{\bigcup}} \Omega_{l_s} \\ \ell^{k_s}_{l_s}, & \text{if } X \in \Omega_{l_s}, \, l_s = 1, \cdots, I \end{cases} \qquad (2.115)$$

Secondly, we must define the derivatives $\frac{\partial U_m}{\partial x_l}$ and $\frac{\partial U_m}{\partial y_l}$. For that purpose consider that all space nodes are fixed except the node $X_l = (x_l, y_l)$. Thus U_m will change only in the support of this node. Let Ω_i be one of the finite elements defining the node support and $X \in \Omega$. Then from 2.98,

$$\frac{\partial U_{m,i}}{\partial x_l} = \sum_{k=1}^{p} \frac{\partial \ell^k_i}{\partial x_l}(X) U^k_{m,i} \qquad (2.116)$$

and

$$\frac{\partial U_{m,i}}{\partial y_l} = \sum_{k=1}^{p} \frac{\partial \ell^k_i}{\partial y_l}(X) U^k_{m,i} \qquad (2.117)$$

From 2.99

$$\frac{\partial \ell^k_i}{\partial x_l} = \sum_{N=1}^{3} \frac{\partial}{\partial L_N}\left(B^{(k)}(L_N)\right)\frac{\partial L_N}{\partial x_l} \prod_{\substack{K=1 \\ K \neq N}}^{3} B^{(r)}(L_K) \qquad (2.118)$$

and

$$\frac{\partial \ell_i^k}{\partial y_l} = \sum_{N=1}^3 \frac{\partial}{\partial L_N}\left(B^{(k)}\left(L_N\right)\right)\frac{\partial L_N}{\partial y_l} \prod_{\substack{K=1\\K\neq N}}^3 B^{(r)}(L_K) \tag{2.119}$$

Assuming that X_l is the K_1 th-vertex of triangle Ω_i, then combining the implicit derivatives of 2.89 with respect to x and to x_{K_1}, and recalling that $L_1 + L_2 + L_3 = 1$ and $|\mathcal{D}| \neq 0$, yields

$$\frac{\partial L_N}{\partial x_{K_1}} = -\frac{\partial L_N}{\partial x}L_{K_1} \tag{2.120}$$

and in a similar way,

$$\frac{\partial L_N}{\partial y_{K_1}} = -\frac{\partial L_N}{\partial y}L_{K1} \tag{2.121}$$

Inserting Equation 2.120 into Equation 2.118 and comparing with Equation 2.103 it follows that:

$$\frac{\partial \ell_i^k}{\partial x_l} = -L_{K1}\frac{\partial \ell_i^k}{\partial x} \tag{2.122}$$

Proceeding similarly, if we insert Equation 2.121 into Equation 2.119 and compare with Equation 2.104

$$\frac{\partial \ell_i^k}{\partial y_l} = -L_{K1}\frac{\partial \ell_i^k}{\partial y} \tag{2.123}$$

Putting together Equation 2.116, Equation 2.122, and Equation 2.101 we obtain

$$\frac{\partial U_{m,i}}{\partial x_l} = -L_{K1}\frac{\partial U_{m,i}}{\partial x} \tag{2.124}$$

And by Equation 2.117, Equation 2.122, and Equation 2.102,

$$\frac{\partial U_{m,i}}{\partial y_l} = -L_{K1}\frac{\partial U_{m,i}}{\partial y} \tag{2.125}$$

We conclude that Equation 2.112, Equation 2.124, and Equation 2.125 define the derivatives of the approximate solution U_m with respect to the mesh deformation allowing the establishment of the governing equations.

We follow the procedure described in Section 2.2 leading to a system of ordinary differential equations for node positions X_l and nodal amplitudes $U_{m,i}^k$, by minimizing

$$\mathcal{T} = \sum_{m_e=1}^n \iint_\Omega (R_{m_e})^2 \, d\Omega + \sum_{j_e=1}^q \left(\epsilon_{j_e}\frac{dZ_{j_e}}{dt} - S_{j_e}\right)^2 \tag{2.126}$$

with respect to the time derivatives

$$\begin{aligned}
\dot{U}_{m,i}^k, \quad m &= 1,...,n; \; i=1,...,q; \; k=1,...,p \\
\dot{x}_l, \dot{y}_l, \quad l &= 1,...,n_s
\end{aligned} \tag{2.127}$$

The general penalty functions are given by Equation 2.40 and R_{m_e} is the residual of the m_e-*th* PDE defined as in 1D case. This minimizing process leads to

$$\begin{cases} \dfrac{\partial T}{\partial \dot{U}^k_{m,i}} = 0 \\ \dfrac{\partial T}{\partial \dot{x}_l} = 0 \wedge \dfrac{\partial T}{\partial \dot{y}_l} = 0 \end{cases} \tag{2.128}$$

First equation The first equation of the system defined by Equation 2.128 is equivalent to

$$\iint_\Omega R_m \frac{\partial U_m}{\partial U^k_{m,i}} d\Omega = 0 \tag{2.129}$$

for $m = 1, ..., n$, $i = 1, ..., q$ and $k = 1, ..., p$.

Let s_i^k be the local node associated with global node Δ_j with support Ω^j. Then using the definition of the residual PDE and the results given by Equation 2.112, the first equation defined by Equation 2.128 is of the form

$$\iint_{\Omega^j} \Phi_j \frac{\partial U_m}{\partial t} d\Omega = \iint_{\Omega^j} \left[\Phi_j \sum_{l_e=1}^n f_{m,l_e} \frac{\partial^2 U_{l_e}}{\partial x^2} \right] d\Omega$$

$$+ \iint_{\Omega^j} \left[\Phi_j \sum_{l_e=1}^n g_{m,l_e} \frac{\partial^2 U_{l_e}}{\partial y^2} \right] d\Omega \tag{2.130}$$

$$+ \iint_{\Omega^j} (\Phi_j h_m) \, d\Omega$$

for $m = 1, ..., n$ and $j = 1, ..., D$.

Second equation Suppose that Δ_j is the *l-th* spacial node $X_l = (x_l, y_l)$. Let I be the number of triangles on the support of Δ_j, and $\dot{A}(\Omega_{i_s})$ the time derivative of the area element Ω_{i_s}. Assume also that X_l is the K_{1_s} vertex on the finite element Ω_{i_s}. Thus the second equation of the system defined by Equation 2.128 is equivalent to

$$\iint_{\Omega^j} \sum_{m_e=1}^n \frac{\partial U_{m_e}}{\partial x_l} \frac{\partial U_{m_e}}{\partial t} d\Omega = \iint_{\Omega^j} \left(\sum_{m_e=1}^n \frac{\partial U_{m_e}}{\partial x_l} \sum_{j_e=1}^n f_{m_e,j_e} \frac{\partial^2 U_{j_e}}{\partial x^2} \right) d\Omega$$

$$+ \iint_{\Omega^j} \left(\sum_{m_e=1}^n \frac{\partial U_{m_e}}{\partial x_l} \sum_{j_e=1}^n g_{m_e,j_e} \frac{\partial^2 U_{j_e}}{\partial y^2} \right) d\Omega$$

$$+ \iint_{\Omega^j} \left(\sum_{m_e=1}^n \frac{\partial U_{m_e}}{\partial x_l} h_{m_e} \right) d\Omega$$

$$- \sum_{s=1}^I \frac{1}{2} \epsilon_{i_s} \Delta y_{K_{1_s}} \left(\epsilon_{i_s} \dot{A}(\Omega_{i_s}) - S_{i_s} \right) \tag{2.131}$$

for $l = 1, ..., n_s$.

Third equation Similarly, the third equation is

$$
\iint_{\Omega^j} \sum_{m_e=1}^{n} \frac{\partial U_{m_e}}{\partial y_l} \frac{\partial U_{m_e}}{\partial t} d\Omega = \iint_{\Omega^j} \left(\sum_{m_e=1}^{n} \frac{\partial U_{m_e}}{\partial y_l} \sum_{j_e=1}^{n} f_{m_e,j_e} \frac{\partial^2 U_{j_e}}{\partial y^2} \right) d\Omega
$$

$$
+ \iint_{\Omega^j} \left(\sum_{m_e=1}^{n} \frac{\partial U_{m_e}}{\partial y_l} \sum_{j_e=1}^{n} g_{m_e,j_e} \frac{\partial^2 U_{j_e}}{\partial y^2} \right) d\Omega
$$

$$
+ \iint_{\Omega^j} \left(\sum_{m_e=1}^{n} \frac{\partial U_{m_e}}{\partial y_l} h_{m_e} \right) d\Omega
$$

$$
- \sum_{s=1}^{I} \frac{1}{2} \epsilon_{i_s} \Delta x_{K_{1_s}} \left(\epsilon_{i_s} \dot{A}(\Omega_{i_s}) - S_{i_s} \right) \quad (2.132)
$$

for $l = 1, ..., n_s$.

All integrals appearing on Equation 2.130, Equation 2.131, and Equation 2.132 must be evaluated over the support of a global node Δ_i. According to the 1D approach, here we choose a smoothing procedure in order that all integrals may be evaluated either analytically or by a suitable quadrature depending on the complexity of the integrand function. In particular, those involving only products of the basis functions are calculated analytically making use of natural coordinates and using the integration by parts [54]. If the node is a vertex or lies on an edge, we shall consider a smoothed function on the neighborhood of a vertex or on the neighborhood of adjacent edges with sufficient continuity to ensure that all integrals exist and therefore may be numerically evaluated.

Motivated by the 1D procedure, the smoothed function of U_m, $\widetilde{U}_j$, is thought such as its first spatial derivatives are continuous and second spatial derivatives exist across element edges. Consider a global node Δ_i and suppose that node is a vertex of a triangle. We shall decompose the support Ω^i of node Δ_i by considering its interior and a δ-neighborhood, V_δ, as Figure 2.10 shows.

Over the interior of Ω^i we consider Gaussian quadrature formulas for triangles [55]. The integrals over V_δ, when considering the opposite δ-edges to Δ_i vanish when $\delta \to 0$ because $\Phi_i \sim \mathcal{O}(\delta)$. With no loss of accuracy we also neglected the integrals over $V\delta$, (the six small δ-triangles around Δ_i) as Figure 2.10 shows. We next have to investigate how to compute the integrals over the δ-neighborhood of adjacent edges of global node Δ_i. For simplicity of the notation let E_δ be the δ-neighborhood of adjacent edges of global node Δ_i. We need to evaluate the integral of the form,

$$
\mathcal{I}_a = \iint_{E_\delta} \left[\mathcal{F}_{m_e} \left(f_{m_e,j_e} \frac{\partial^2 \widetilde{U}_{j_e}}{\partial x^2} + g_{m_e,j_e} \frac{\partial^2 \widetilde{U}_{j_e}}{\partial y^2} \right) \right] dxdy \quad (2.133)
$$

where a is the length of the edge and $\mathcal{F}_{m_e}$ is a global basis function or a sum involving first-order spatial derivatives. The integral $\mathcal{I}_a$ is evaluating by

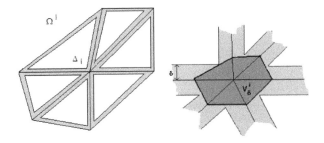

FIGURE 2.10
V_δ,δ-neighborhood over the support Ω^i (left) and V_δ^i, δ-neighborhood around spatial nodeΔ_i (right). Adapted with permission from Coimbra et al. [42]. ©2003, Elsevier.

considering the new coordinate system (ξ, η) where η is the axis along the edge as Figure 2.11 illustrates.

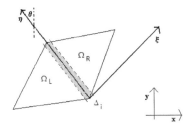

FIGURE 2.11
The δ-neighborhood of an adjacent edge to global node Δ_i. Adapted with permission from Coimbra et al. [42]. ©2003, Elsevier.

Thus the integral defined by Equation 2.133 reduces to

$$\mathcal{I}_a = \int_0^a \left[\int_{-\delta}^{\delta} \mathcal{F}_{m_e} \left(f_{m_e,j_e} \cos^2\theta + g_{m_e,j_e} \sin^2\theta \right) \frac{\partial^2 \tilde{U}_{j_e}}{\partial \xi^2} d\xi \right] d\eta \qquad (2.134)$$

where the (ξ, η) dependence is omitted for simplification of the notation. Furthermore, following the idea of the smoothing approach, the integral defined by Equation 2.134 can be evaluated by some quadrature rule. Then, taking $\delta \to 0$, yields

$$\mathcal{I}_a = a \sum_Q w(Q)\mathcal{E}_{m_e}(Q) \left(\frac{\partial U_{j_e}}{\partial \xi}\bigg|_{Q\in\Omega_R} - \frac{\partial U_{j_e}}{\partial \xi}\bigg|_{Q\in\Omega_L} \right)$$

where $w(Q)$ is the 1D weight quadrature associated to the quadrature point Q and

$$\mathcal{E}_{m_e}(Q) = \mathcal{F}_{m_e} \left(f_{m_e,j_e} \cos^2 \theta + g_{m_e,j_e} \sin^2 \theta \right) \Big|_Q \qquad (2.135)$$

The right and the left triangles with common oriented edge are, respectively, Ω_R and Ω_L as Figure 2.11 shows. The details of this technique can be found in [54, 42].

All the above theory concerning the integrals computation makes possible the establishment of the governing equations. According to the prescribed boundary conditions some equations may have to be revised and rewritten. The problem then requires the solution of a large system of ordinary differential equations. The MFEM takes advantage of the available of high-quality ODE solvers in order to determine the unknown nodal positions and nodal amplitudes. In the 2D-MFEM numerical code we use the LSODIS package [51] but others solvers are available, [48, 50, 49].

2.3.2   MFEM formulation for two scales: 1D+1d models

In this section, a numerical algorithm based on the MFEM is developed to solve time-dependent partial differential equations defined in a two-dimensional spatial domain with a specific structure [44].

The fundamental concepts of the moving finite elements method were developed in Section 2.2. For 1D time-dependent partial differential equations, we consider the use of higher-order Lagrange basis functions for both the trial and the test spaces. Furthermore, the use of different spatial meshes for each dependent variable is admissible. As shown in [56], the MFEM produces efficient and accurate numerical solutions for real problems in one or two space dimensions. Here, we consider a large class of two-dimensional problems, where the presence of two different spatial scales, a macroscale and a microscale can be observed. For this kind of problem, we developed a numerical algorithm that extends the existing knowledge of the 1D moving finite elements method.

The basic idea behind this new numerical algorithm, called 1D+1d MFEM, is quite straightforward. First, the time-dependent macro equation is discretized in space using the 1D-MFEM. Second, to each global node of the macro spatial mesh we associate a micro problem, which is also discretized in space using the 1D-MFEM. Figure 2.12 shows an example of a macro spatial mesh with 3 finite elements and global nodes corresponding to a piecewise polynomial of degree 2 in the first element, degree 3 in the second element, and a linear approximation in the third element. In this example the global mesh has 7 global nodes, and thus 7 micro problems must be considered.

A crucial issue is to establish the links between the two scales. Our strategy is to include the link between the two scales in the equation associated with the macro problem and in the boundary condition at the right endpoint of

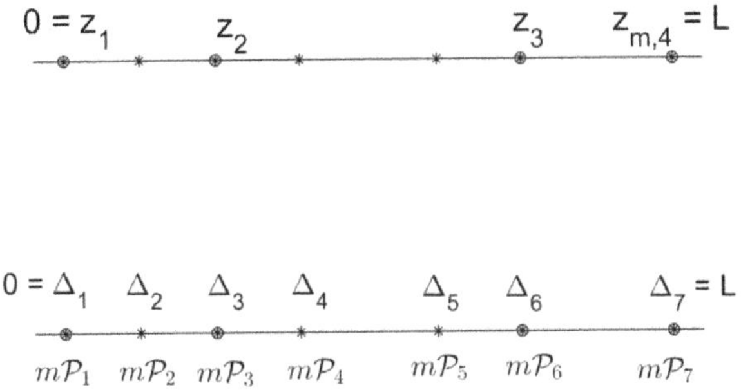

FIGURE 2.12
Micro problems, $mP_i, i = 1, \cdots, 7$ associated to a macro spatial mesh with 7 global nodes.

the spatial domain of each micro problem [44]. Finally the resulting system of ordinary differential equations must be solved to get the numerical solution.

Formulation

Let us now consider the 1D+1d model described by a 1D macro equation coupled with several micro equations. Let z, $0 \leq z \leq L$, be the spatial coordinate for the macro equation and r, $0 \leq r \leq 1$, the normalized spatial coordinate for the micro equations. The time variable is $t \geq 0$.

The dependent variable for the macro equation is $u = u(z,t)$. The dependent variable for the micro equations is represented by $u_p = u_p(r, z, t)$. We shall assume that the general form for the macro equation is given by

$$\frac{\partial u}{\partial t} = f\left(z, t, u, \frac{\partial u}{\partial z}\right) \frac{\partial^2 u}{\partial z^2} + h\left(z, t, u, \frac{\partial u}{\partial z}\right) + \vartheta\left(u_p(1, z, t)\right) \qquad (2.136)$$

where the term ϑ expresses the link term between the macro- and microscales. The diffusive coefficient is defined by f. The h term includes convection and reaction phenomena. Without ϑ this equation is an explicit equation of the form 2.6.

The micro equations are assumed to be of the form

$$\frac{\partial u_p}{\partial t} = \bar{f}\left(r, t, u_p, \frac{\partial u_p}{\partial r}\right) \frac{\partial^2 u_p}{\partial r^2} + \bar{h}\left(r, t, u_p, \frac{\partial u_p}{\partial r}\right), \qquad (2.137)$$

where $\bar{f}$ is the diffusive term and $\bar{h}$ includes convection and reaction. This kind of equation is an explicit equation of the form 2.6. Initial conditions for u and u_p must be given to start the problem. Together with Equation 2.136,

Dirichlet, Neumann, or Robin boundary conditions may be considered. For the micro equations, Dirichlet, Neumann, or Robin boundary conditions are accepted for the left endpoint of spatial domain, $r = 0$. However for $r = 1$ we impose the following boundary condition with a term ϖ that links both scales:

$$\frac{\partial u_p}{\partial r}(1, z, t) = \alpha\, u_p + \beta + \varpi\Big(u(z, t)\Big) \qquad (2.138)$$

Using the previous notations we can say that the link between both scales is done by the functions ϑ and ϖ.

Space discretization

The spatial discretization is performed in two steps. First, the discretization process starts with the macro domain defining an initial mesh by using q finite elements.

$$\mathcal{M}: \quad a \leq z_1(t) < \ldots < z_j(t) < \ldots < z_{q+1}(t) \leq b. \qquad (2.139)$$

In each one of them we use Lagrange interpolation to find a piecewise polynomial approximate solution, U, taking a collection of p_j local points in each finite element. The next step is the discretization of the micro domains by using 1D-MFEM. Micro problems are associated with the global nodes of macro mesh $\mathcal{M}$. Therefore the z position of the micro problems also varies with time. For example, if the position z_0 of the macro domain is associated with a micro problem at time t, then the dependent variable for this micro problem is $u_p = u_p(r, z_0, t)$. However, it is possible that for a different time, at the same position z_0 of the macro domain, there is any micro problem associated. What happens then is that for a given instant t we have a collection of values $\{u_p(1, \Delta_i, t)\}$ which we use to calculate $u_p(1, z, t)$ at a generic point z of the macro domain using Lagrange interpolation. Therefore, it is important that global nodes Δ_i are well distributed along the macro domain.

We discretize the micro domains with different meshes. Note that there is no explicit link between micro problems. However, from the numerical viewpoint, the link between both scales is an important issue of our scheme. Consequently ϑ and ϖ terms should be treated with particular caution.

Due to the strategy adopted for dealing with 1D+1d models, in general terms, we can adopt the governing equations given by Equation 2.81, Equation 2.82, and Equation 2.83 with minor changes. For calculations concerning the macro domain equation the term $F_{m,j}^k$ given by Equation 2.61 should be rewritten to incorporate the new function ϑ. In fact we have $m = 1$ and $F_{1,j}^k$ is now given by

$$F_{1,j}^k = \frac{1}{Z_{1,j}} \sum_{i=1}^{n} \int_{z_j^+}^{z_{j+1}^-} \ell_{1,j}^k(x)\Big(f\frac{\partial^2 U}{\partial z^2} + h + \vartheta\Big)dx. \qquad (2.140)$$

For the micro equations the exterior boundary condition at $r = 1$ will be given by Equation 2.138, where u_p is replaced by the value of the approximate solution U_p calculated at $r = 1$, and ϑ is to be evaluated as we now explain.

If we are discretizing a micro equation associated with some global node Δ_i in the macro domain and we want to implement the boundary condition at $r = 1$, we first determine $U(\Delta_i, t)$ using Lagrange interpolation in the finite element of the macro domain that contains Δ_i and then we calculate the term $\varpi\left(U(\Delta_i, t)\right)$ to completely define Equation 2.138.

On the other hand, to calculate the residual of the macro equation at some point z^* of the macro domain, we only have a set of approximate values $\{U_p(1, \Delta_i, t)\}$. By using Lagrange interpolation we can find the approximate solution $U_p(1, z^*, t)$ which are used to determine $\vartheta\left(U_p(1, z^*, t)\right)$. From the double discretization, macro domain, and micro domains, it results that a large system of ordinary differential equations in time that is solved using a stiff ODE solver.

2.4 Conclusions

In this chapter, we have presented the development of the moving finite elements method to solve one- and two-dimensional convection–diffusion–reaction problems. The main advantage of this procedure is that for any instant of time, the MFEM is able to determine an accurate solution and the mesh simultaneously. Section 2.1 introduces the class of problems we want to solve. In particular we focus our attention on problems involving fine-scale phenomena such as steep moving fronts. Section 2.2 of the chapter provides a detailed development of the formulation of the moving finite elements method with high-order Lagrange basis functions. The fundamental properties of the moving finite elements method in the one-dimensional spatial domain has been analyzed. Questions concerning the singularity of the matrix and the construction of the smoothing approximate solution in a neighborhood of a spatial node were treated. The general equations of the method were deducted. The MFEM generalizes naturally to higher space dimensions and in Section 2.3 two different methodologies are investigated. The first one is the straightforward extension of the 1D-MFEM to deal with evolutionary problems in the two-dimensional space domain. The second one has been outlined to solve real-life processes with two different space scales, a macroscale and a microscale connected by one of the boundaries of the microscale.

Bibliography

[1] R. Byron Bird, Warren E. Stewart, and Edwin N. Lightfoot. *Transport Phenomena*. New York: John Wiley, 2002.

[2] Kuzman Adzievski and Abul Hasan Siddiqi. *Introduction to Partial Differential Equations for Scientists and Engineers Using Mathematica*. CRC Press, 2013.

[3] Bruce A. Finlayson. *Numerical Methods for Problems with Moving Fronts*. Ravenna Park Publishing, 1992.

[4] J. Crank. *The Mathematics of Diffusion*. Oxford University Press, 1979.

[5] C. Zoppou and J. H. Knight. Analytical solution of a spatially variable coefficient advection-diffusion equation in up to three dimensions. *Applied Mathematical Modelling*, 23(9):667–685, 1999.

[6] Kaveh Zamani and Fabián A. Bombardelli. Analytical solutions of nonlinear and variable-parameter transport equations for verification of numerical solvers. *Environmental Fluid Mechanics*, 14(4):711–742, 2014.

[7] M. D. Djomehri, S. Doss, R. Gelinas, and K. Miller. Applications of the moving finite element for systems. *Journal of Computational Physics*, pages 1–41, 1985.

[8] C. Sereno, A. E. Rodrigues, and J. Villadsen. Solution of partial differential equations systems by the moving finite element method. *Computers and Chemical Engineering*, 16(6):583–592, 1992. An International Journal of Computer Applications in Chemical Engineering.

[9] Maria do Carmo Coimbra, C. Sereno, and A. E. Rodrigues. Modelling multicomponent adsorption process by a moving finite element method. *Journal of Computational and Applied Mathematics*, 115(1-2):169–179, 2000.

[10] Ne-Zheng Sun. *Mathematical Modeling of Groundwater Pollution*. Springer, 1996.

[11] J. D. Murray. Multi-species waves and practical applications. In J. D. Murray, editor, *Mathematical Biology*, volume 18 of *Interdisciplinary Applied Mathematics*, pages 1–70. Springer New York, 2003.

[12] Graham W. Griffiths and William E. Schiesser. *Traveling Wave Analysis of Partial Differential Equations*. Academic Press, 2012.

[13] W. E. Schiesser. *Differential Equation Analysis in Biomedical Science and Engineering: Partial Differential Equation Applications with R*. Wiley, 2014.

[14] William F. Ames. *Numerical Methods for Partial Differential Equations.* Academic Press, 1977.

[15] Bruce A. Finlayson. *Nonlinear Analysis in Chemical Engineering.* McGraw-Hill, 1980.

[16] Williem Hundsdorfer and Jan Verwer. *Numerical Solution of Time-Dependent Advection-Diffusion-Reaction Equations.* Springer, 2003.

[17] G. Strang and G. J. Fix. *An Analysis of the Finite Element Method.* Prentice-Hall, 1973.

[18] Mikhail Shashkov. *Conservative Finite-Difference Methods on General Grids.* CRC Press, 1995.

[19] Leon Lapidus and George F. Pinder. *Numerical Solution of Partial Differential Equations in Science and Engineering.* Wiley, 1999.

[20] A. Richard Mitchell and D. F. Griffiths. *The Finite Difference Method in Partial Differential Equations.* Wiley, 1980.

[21] Chandrakant S. Desai and Tribikram Kundu. *Introductory Finite Element Method.* CRC Press, 2001.

[22] H. C. Martin and G. F. Carey. *Introduction to Finite Element Analysis.* McGraw-Hill, 1973.

[23] P. A. Zegeling. *Moving-Grid Methods For Time-Dependent Partial Differential Equations*, volume 94 of *CWI Tracts.* CWI, 1993.

[24] M. J. Baines. *Moving Finite Elements.* Clarendon Press, 1994.

[25] Weizhang Huang and Robert D. Russell. *Adaptive Moving Mesh Methods.* Springer-Verlag New York, 2011.

[26] D. F. Hawken, J.J. Gottlieb, and J.S. Hansen. Review of some adaptive node-movement techniques in finite-element and finite-difference solutions of partial differential equations. *Journal of Computational Physics*, 95(2):254–302, 1991.

[27] Paul A. Zegeling. Moving grid techniques. In Nigel P. Weatherill, Bharat K. Soni, and Joe F. Thompson, editors, *Handbook of Grid Generation.* CRC Press, 1998.

[28] Weiming Cao, Weizhang Huang, and Robert D. Russell. Approaches for generating moving adaptive meshes: Location versus velocity. *Applied Numerical Mathematics*, 47(2):121–138, 2003.

[29] James M. Hyman. *The Method of Lines Solution of Partial Differential Equations.* Courant Institute of Mathematical Sciences, New York University, 1976.

[30] W. E. Schiesser. *The Numerical Method of Lines.* Academic Press, 1991.

[31] Alain Vande Wouwer, Philippe Saucez, and William Schiesser. *Adaptive Method of Lines.* Chapman and Hall/CRC, 2001.

[32] K. Miller and R. Miller. Moving finite elements i. *SIAM Journal on Numerical Analysis*, 18(6):1019–1032, 1981.

[33] K. Miller. Moving finite elements ii. *SIAM Journal on Numerical Analysis*, 18(6):1033–1057, 1981.

[34] K. Miller. A geometrical-mechanical interpretation of gradient-weighted moving finite elements. *SIAM Journal on Numerical Analysis*, 34(1):67–90, 1997.

[35] N. N. Carlson and K. Miller. Design and application of a gradient-weighted moving finite element code i: In one dimension. *SIAM Journal on Scientific Computing*, 19(3):728–765, 1998.

[36] N. N. Carlson and K. Miller. Design and application of a gradient-weighted moving finite element code ii: In two dimensions. *SIAM Journal on Scientific Computing*, 19(3):766–798, 1998.

[37] Keith Miller and Mike J. Baines. Least squares moving finite elements. *IMA Journal of Numerical Analysis*, 21(3):621–642, 2001.

[38] Keith Miller. Stabilized moving finite elements for convection dominated problems. *Journal of Scientific Computing*, 24(2):163–182, 2005.

[39] B. M. Herbst, S. W. Schoombie, and A. R. Mitchell. A moving Petrov–Galerkin method for transport equations. *International Journal for Numerical Methods in Engineering*, 18:1321–1336, 1982.

[40] J. A. Hanser and O. Hassager. A new moving finite element method based on quadratic approximation functions. *Int. J. Numer. Meth. Engn.*, 28:415–430, 1989.

[41] C. Sereno, A. E. Rodrigues, and J. Villadsen. The moving finite element method with polynomial approximation of any degree. *Computers and Chemical Engineering*, 15(1):25–33, 1991.

[42] Maria do Carmo Coimbra, C. Sereno, and A. E. Rodrigues. A moving finite element method for the solution of two-dimensional time-dependent models. *Applied Numerical Mathematics*, 44(4):449–469, 2003.

[43] Belmiro P. M. Duarte and Cristina M. S. G. Baptista. Moving finite elements method applied to dynamic population balance equations. *AIChE Journal*, 54(3):673–692, 2008.

[44] J. D. Rodrigues. *Desenvolvimento do Método de Elementos Finitos Móveis para a simulação de processos*. PhD thesis, Universidade do Porto, 2010.

[45] R. J. Robalo. *O Método de Elementos Finitos Móveis para problemas evolutivos com Fronteira Móvel*. PhD thesis, Universidade da Beira Interior, 2014.

[46] C. Sereno. *Método dos Elementos Finitos Móveis: Aplicações em Engenharia Química*. PhD thesis, Universidade do Porto, 1990.

[47] R. J. Gelinas, S. K. Doss, and Keith Miller. The moving finite element method: Applications to general partial differential equations with multiple large gradients. *Journal of Computational Physics*, 40(1):202–249, 1981.

[48] L. R. Petzold. A description of dassl: A differential-algebraic system solver. In R. S. Stepleman and et al., editors, *IMACS Trans. Scientific Computing*, page 65. North-Holland, 1983.

[49] L. F. Shampine. Ode solvers and the method of lines. *Numerical Methods for Partial Differential Equations*, 10:739–755, 1994.

[50] E. Hairer and G. Wanner. Stiff differential equations solved by radau methods. *Journal of Computational and Applied Mathematics*, 111(1-2):93–111, 1999.

[51] A. C. Hindmarsh. LSODE and LSODIS, two new initial value ordinary differential equation solvers. *ACM-Signum Newslett*, 15:10–11, 1980.

[52] A. C. Hindmarsh. ODEPACK, a systematized collection of ode solvers. In R. S. Stepleman et al., editors, *Scientific Computing*, pages 55–64. Elsevier, 1983.

[53] T. J. Chung. *Finite Element Analysis in Fluid Dynamics*. McGraw-Hill, 1978.

[54] Maria do Carmo Coimbra. *Método dos Elementos Finitos Móveis: Aplicação a sistemas de equações de derivadas parciais bidimensionais*. PhD thesis, Universidade do Porto, 2000.

[55] G. R. Cowper. Gaussian quadrature formulas for triangles. *International Journal for Numerical Methods in Engineering*, 7:405–408, 1973.

[56] Maria do Carmo Coimbra, C. Sereno, and A. E. Rodrigues. Moving finite element method: Applications to science and engineering problems. *Computers and Chemical Engineering*, 28(5):597–603, 2004.

3

Solving 1D Time-Dependent Models

3.1 Introduction, Description, and Implementation of MFEM Code

The primary focus of this third chapter is the numerical moving finite element method solutions of complex real-world time-dependent partial differential problems in one spatial dimension. The examples illustrate the working of the MFEM, some of its difficulties, and the techniques proposed to overcome them. The MFEM algorithm is implemented in FORTRAN language. There are many different FORTRAN compilers that can be used and a lot of scientific code is written in FORTRAN language [1]. The computations for this chapter were done on an Intel® Core™ i7-4790 @ 3.60GHz personal computer.

User supply functions and data

MFEM computer code is a set of FORTRAN routines and functions for the numerical integration of Equation 2.6,

$$\mathbf{G}\left(x, t, \mathbf{u}\right) \frac{\partial \mathbf{u}}{\partial t} = \mathbf{F}\left(x, t, \mathbf{u}, \mathbf{u}_x\right) \frac{\partial^2 \mathbf{u}}{\partial x^2} + \mathbf{h}\left(x, t, \mathbf{u}, \mathbf{u}_x\right), a \le x \le b, \ t > 0$$

under certain initial conditions and boundary conditions.

The functions $\mathbf{G}$, $\mathbf{F}$, and $\mathbf{h}$ must be supplied by the user as FORTRAN functions of the form, respectively,

---

Model Equations.

```
DOUBLE PRECISION FUNCTION F3(M,I,X,T,U)
DOUBLE PRECISION FUNCTION F1(M,I,X,T,U,DUDX)
DOUBLE PRECISION FUNCTION F2(M,I,X,T,U,DUDX)
```

The type of boundary conditions is specified in the data file for MFEM code. Besides, two functions must be declared to define the boundary conditions, BC1 associated to the left end of the spatial domain and BC2 defining the boundary conditions at the right end of the spatial domain.

Boundary conditions functions.

```
DOUBLE PRECISION FUNCTION BC1(M,U)
DOUBLE PRECISION FUNCTION BC2(M,U)
```

A function defining the initial conditions is also needed,

Initial condition function.

```
DOUBLE PRECISION FUNCTION IC(M,U)
```

As usual, the user of a numerical code should provide the system with a set of parameters defining the problem to be solved and additionally parameters concerning the MFEM and the user choices. In fact, the use of the MFEM code requires the definition of some specific parameters. The system to be solved may contain ordinary differential equations besides the partial differential equations. The user must first indicate the number of ODEs and then the number of PDEs to solve. It is necessary to indicate the spatial domain, Ω. Unless specified, both endpoints of Ω are considered to be local nodes. The user may choose the number of quadrature points, the number of finite elements, and the type of boundary conditions must be defined.

The degree of the approximation in each finite element is defined by indicating the number of interior local nodes in each finite element. By default, we use the same number of interior local nodes in all finite elements. The value of the small constants to define penalty functions, the relative and absolute tolerances for the ODE solver, and the starting space grid for each dependent variable must also be provided. The number of quadrature points must be equal to or higher than the largest number of interior local nodes used and by default 11 quadrature points are considered. The relative and absolute tolerance for the ODE solver are usually set to $tol_1 = tol_2 = 10^{-5}$, but smaller tolerances may be required.

Concerning the user supply small penalty constantsave needed to define the penalty functions $S_{m,j}$ and $\epsilon_{m,j}$ given by Equation 2.40. It is important to note that we can use the same set of values in all finite elements or make some adjustments, element by element, if necessary. $c_5^{m,j}$ is the permissible minimum distance between the two consecutive nodes $X_{m,j}$ and $X_{m,j+1}$ of the spatial mesh $\mathcal{M}_m$. The constants $c_1^{m,j}$ and $c_2^{m,j}$ enter in the definition of the spring forces between spatial nodes. In our numerical simulations we consider $c_2^{m,j} = 0$ and $c_5^{m,j} = c_6^{m,j} = 10^{-5}$ for all finite elements. With this choice, we verify that in increasing the value of the constant $c_1^{m,j}$ the adaptivity of the mesh is reduced. So we start using a small value, for example, the same value defined for the minimum nodal distance. Constants $c_3^{m,j}$ and $c_4^{m,j}$ relate to the viscosity function influencing the nodes movement and our choice in most cases is to set $c_3^{m,j} = 10^{-1} c_4^{m,j}$. By default we consider the same set of penalty constants for all finite elements and we recommend the following set of values

$$c_1^{m,j} = 10^{-5}, \ c_2^{m,j} = 0, \ c_3^{m,j} = 10^{-3}, \ c_4^{m,j} = 10^{-2}, \ c_5^{m,j} = c_6^{m,j} = 10^{-5}$$

When the solution profile at the initial time contains large gradients, the choice of the initial meshes deserves particular attention. For smooth initial profiles, we start with an equally spaced mesh, otherwise initial nodes are placed to give a good representation of the initial profile. In some cases, we can take advantage of the knowledge we have about the problem to put the initial spatial nodes in regions where large gradients are expected.

All these data must be specified in a file with the structure presented below.

Data file for MFEM code. Part I

```
Number of ODE, Number of PDE
{Space Domain:} X_left,X_right
Number of quadrature points
{mesh 1:}
Number of finite elements,Type of boundary conditions
Number of interior local nodes in each finite element
Set of penalty constants for first finite element
...
Set of penalty constants for last finite element
...
{mesh n:}
Number of finite elements,Type of boundary conditions
....
Set of penalty constants for last finite element
Relative tolerance, Absolute tolerance for ODE solver
Starting space mesh 1
...
Starting space mesh n
```

MFEM computer code

To start it is required to consider a set of routines to define all the values in the reference element, $[0, 1]$, needed for interpolation and integration. For that purpose, we use the subroutines JCOBI, DFOPR, RADAU, and INTRP, developed by Villadsen and Michelsen [2].

These routines together with the initial nodes positions provide the initial global meshes, and also the initial solution on those meshes must be computed.

The implicit system of ODE that results from the application of the MFEM is solved by using the ODEPACK package [3,4], and the construction of the ODE system is the central task of the MFEM computer code. The ODE-PACK solver for ordinary differential equations, LSODIS, solves the initial value problem for linearly implicit systems of first order,

$$a(t, y) * dy/dt = g(t, y) \qquad (3.1)$$

Thus, it is required to initialize the Y ODE vector. The routine CALC_NOS_INT defines the initial global nodes and the initial solution at each global node as well as the LSODIS vector Y.

Initialization of ODE vector.

```
CALL CALC_NOS_INT (NY,Y,MAXVP,NVIN,XOUT1,YOUT1,VIN)
```

The communication between the MFEM and the LSODIS package is done by the routines RES, ADDA, and JAC. The routine RES computes the residual function, $r = g(t, y) - a(t, y) * dy/dt$, as a function of t and the vectors y and dy/dt. The routine ADDA is used for computing the matrix $a(t, y)$. If the Jacobian matrix dr/dy is being supplied, we must provide the subroutine JAC which computes dr/dy as a function of t, y, and dy/dt. MFEM equations described in Section 2.2 will be implemented to define these routines.

Selected numerical output

The presentation of results is of great importance and, the way they can be easily interpreted is an added value. The scientific visualization is widespread in most applications and thus many software packages have been developed to make this task easy and simple. All figures presented here are produced in MATLAB® [5] environment taking advantage of its powerful visualization capabilities. To obtain the data files with the MFEM results, the user has to select the numerical output to be produced. Thus the user data file for running an MFEM problem must also specify the desired details for the output. Two main visualizations can be produced to illustrate the model behavior: axial profiles and histories. An axial profile is a plot of the solution as a function of space variable at a certain time $t = t_i$ while histories are the plots of the solution as a function of time t at a certain space position $x = x_i$. The user may choose both t_i and x_i. The movement of the nodes in time is also registered for plotting the trajectories, i.e., the graphic presenting the position of the spatial nodes as a function of time variable.

Data file for MFEM code. Part II

```
Number of points to plot interpolated solution
Number of axial profiles
Times at which the profiles are plotted
Final time for ODE integration
Number of histories
Axial positions for plot histories
Time output for plot histories
Time output for plot nodes movement
```

Therefore, after a call of LSODIS routine, to assure the numerical output, a routine converting the LSODIS vector Y to the global nodes $X_{m,j}$ and the value of the approximate solution at the global nodes $U_{m,j}^k$ is also required.

Conversion of ODE vector.

```
SUBROUTINE CONV_VAR (NYV,YV)
```

Numerical simulations

For each example appearing in this chapter, we present a table summarizing the MFEM parameters used in the MFEM numerical simulation. The omission of detailed information indicates that the same data are used for all finite elements.

The information concerning the solution of the ODE system, the time interval, and the CPU time are also presented in a table displaying the following information:

- NST, the number of steps taken for the problem over the given time interval.

- NRE, the number of residual evaluations.

- NJE, the number of Jacobian evaluations.

- Time interval, independent variable for ODE integration.

- CPU time, the CPU time required to achieve the integration, expressed in seconds.

3.2   Burgers' Equation

Our first example is one of the most studied equations, the Burgers' equation [6] . We consider the nonlinear dimensionless model where u is the dimensionless unknown function we are looking for, x is the space variable normalized by the characteristic length L, and t is the time variable normalized by $\tau = L/v_{ref}$, where v_{ref}, is the reference velocity, [7],

$$\frac{\partial u}{\partial t} = d\frac{\partial^2 u}{\partial x^2} - \frac{\partial f(u)}{\partial x}, \ 0 \leq x \leq 1, \ t > 0 \qquad (3.2)$$

with initial condition $u(x,0) = u_0(x)$ and assuming Dirichlet boundary conditions. This equation appears often as a simplification of a more complex model in different kinds of applications and a large amount of literature is available concerning Burgers' equation [8, 9, 10, 11]. Since Equation 3.2 combines both nonlinear propagation effects and diffusive effects, it is often taken as a test model for numerical methods. Besides, for some initial and boundary conditions, it possesses analytical solutions, and so the accuracy of numerical solutions can be verified [7, 12, 13, 14]. Decreasing the value of diffusion parameter, d, the solution develops moving shock waves which makes the problem more challenging to numerical methods.

Here we consider $f(u) = u^2/2$. Two different problems are presented by changing the initial condition $u_0(x)$. The MFEM parameters for all numerical simulations are the same (see Table 3.1).

TABLE 3.1
MFEM parameters. Burgers' equation.

Number of finite elements	10
Initial mesh	equally spaced
Approximation degree	6
Degree of quadrature formula	11
Penalty constants	$10^{-5}, 0, 10^{-3}, 10^{-2}, 10^{-5}, 10^{-5}$
Absolute and relative tolerances	
for ODE solver	$10^{-5}, 10^{-5}$

As a first example, case (i), we select the initial condition,

$$u_0(x) = \sin(\pi x) \qquad (3.3)$$

and homogeneous Dirichlet boundary conditions at $x = 0$ and $x = 1$. Two different values for the diffusion coefficient, $d = 10^{-1}$, case (ia) and $d = 10^{-3}$, case (ib) are studied. For the numerical simulation the time interval considered is $0 \leq t \leq 2$.

Let us start with case (ia). For a moderate value of the diffusion coefficient, $d = 10^{-1}$, case (ia), the solution is smooth. MFEM produces an accurate solution. Figure 3.1 (top) shows axial profiles of u for different values of time, $t = 0; 0.2; 0.5; 0.6; 1; 2$.

Even with smooth initial conditions, shocks may occur, if the diffusion parameter is small. It is the case when $d = 10^{-3}$, case (ib). For this value of d, the solution indicates a wave moving to the right that develops a steep front until a layer is formed at $x = 1$. Then the solution slowly decays to zero, keeping the layer at $x = 1$. This is illustrated in Figure 3.1 (bottom) showing the MFEM solution u versus x for different values of time, $t = 0; 0.2; 0.5; 0.6; 1; 2$.

Figure 3.1 shows that for the same model we have solutions with distinct behavior by influence of the diffusion parameter d. Numerical results show the ability of MFEM to produce accurate solutions and to accurately track rapid spatial transitions without the need to adjust MFEM parameters. An important issue is that MFEM is able to put the mesh nodes where they are needed. The movements of nodes position for spatial mesh are shown in Figure 3.2. The initial mesh takes the nodes positions equally spaced over space domain $[0, 1]$. For the smooth solutions, $d = 10^{-1}$, the spacial nodes keep their initial positions (Figure 3.2 (top)) while for small values of diffusion coefficient, $d = 10^{-3}$, nodes maintain their positions until the formation of the steep front and then they move to the right to capture the layer at $x = 1$ (Figure 3.2 (bottom)).

As a second example, case (ii), for Burgers' equation, we set $d = 10^{-3}$ and select $u_0(x)$ and nonhomogeneous Dirichlet boundary conditions matching the exact solution of 3.2,

$$u(x,t) = 1 - 0.9\frac{r_1}{r_1 + r_2 + r_3} - 0.5\frac{r_2}{r_1 + r_2 + r_3} \qquad (3.4)$$

where

$$r_1 = e^{-\frac{20(x - 0.5) + 99t}{400d}}$$

$$r_2 = e^{-\frac{4(x - 0.5) + 3t}{16d}}$$

$$r_3 = e^{-\frac{x - 0.375}{2d}} \qquad (3.5)$$

With these conditions, Burgers' equation has two distinct initial sharp fronts that propagate at different speeds and coalesce to a single propagating front which becomes sharper as time progresses. Spatial interior nodes for the initial mesh are taken equally spaced in the subdomain $[0.2, 0.6]$ that contains the two initial fronts. The numerical simulation considers the time interval $0 \leq t \leq 1$. Figure 3.3 (top) shows the MFEM solution $u(x,t)$ for $t = 0$ up to $t = 1$ with time intervals of 0.1. The positions of the nodes as function of time is presented in Figure 3.3 (bottom). Table 3.2 gives the integration information concerning the numerical simulations.

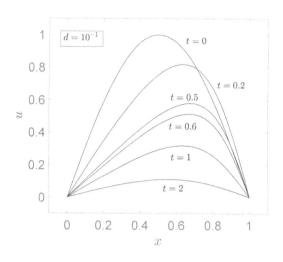

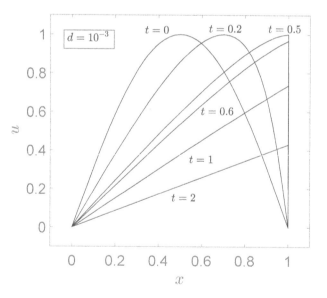

FIGURE 3.1

Burgers' equation, axial profiles of u versus x at $t = 0; 0.2; 0.5; 0.6; 1; 2$ for $d = 10^{-1}$, (top) and $d = 10^{-3}$, (bottom).

3.3   A Boundary Layer Problem

In this section we will consider the relevant linear test model convection–diffusion equation, starting with simple constant coefficients. Let $\nu \in \mathbb{R}$ be the

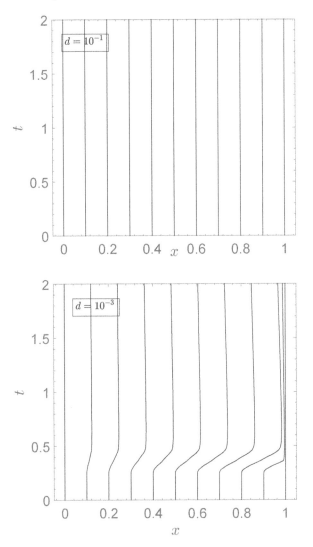

FIGURE 3.2
Burgers' equation, case (i). Trajectories of the nodes with time for $d = 10^{-1}$ (top) and $d = 10^{-3}$ (bottom).

flow velocity and $d > 0$ the diffusion coefficient. Consider the model proposed by Carlson and Miller [15]

$$\frac{\partial u}{\partial t} + \nu \frac{\partial u}{\partial x} = d \frac{\partial^2 u}{\partial x^2}, \qquad (3.6)$$

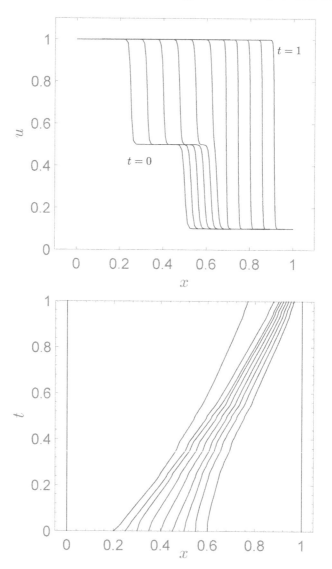

FIGURE 3.3
Burgers' equation, case (ii). Axial profiles of u at $t = 0$ up to $t = 1$ with
time intervals of 0.1 (top) and trajectories of the nodes position with time
(bottom).

with $0 \leq x \leq 1$, and the dimensionless time $t > 0$, with given initial condition

$$u(x,0) = x^{16}, \ 0 \leq x \leq 1 \tag{3.7}$$

TABLE 3.2

Integration information. Burgers' equation.

Simulations	NST	NRE	NJE	CPU time(s)
case (ia)	74	786	10	0.06
case (ib)	268	3262	42	0.25
case (ii)	984	12939	167	0.46

along with Dirichlet boundary conditions,

$$u(0,t) = 0, \ u(1,t) = 1, \ t > 0. \tag{3.8}$$

Let us consider $\nu = 1$ and $d = 10^{-4}$. Since $d << |\nu|$ this problem is convective dominate. We expect to find almost the same numerical behavior as for a pure convection problem. When $d \approx 0$ the Dirichlet condition at the right endpoint of the space domain will give rise to a boundary layer. In fact, the wave traveling to the right boundary with speed 1, reaches $x = 1$, immediately after time $t = 1$ and has a stationary solution with a boundary layer at $x = 1$, [15]. The presence of this boundary layer will require a dynamically adjusting mesh and represents a significant challenge to obtain the numerical solution accurately because the derivatives of the solution increase without bound in the neighborhood of the boundary layer. Similar problems with boundary layer and solutions can be found in [15], [16], and [9]. We solve this problem for $0 \leq t \leq 500$. Table 3.3 summarizes the MFEM parameters used. The MFEM gives a solution of great quality with few nodes as Figure 3.4 shows. We present solutions for $t = 0$ and $t = 500$ and spatial nodes position at these instants. For the starting mesh, $\mathcal{M}_1(t = 0)$, the nodes position is chosen to concentrate the nodes on the left end of the spatial domain, namely

$$X_{1,1} = 0, X_{1,2} = 0.00002, X_{1,3} = 0.0001, X_{1,4} = 0.001, X_{1,5} = 0.01$$
$$X_{1,6} = 0.02, X_{1,7} = 0.03, X_{1,8} = 0.04, X_{1,9} = 1$$

The MFEM enables the nodes to travel from their initial positions near the right end in order to well accommodate the moving wave solution as Figure 3.4 shows. Table 3.4 summarizes the integration information for this simulation.

3.4 A Problem from Mathematical Biology

This third example is concerned with the study of the FitzHugh–Nagumo equations, one of the most widely studied biological systems. These equations arise in mathematical biology as a model of the transmission of ionic current flow impulses across a semi-infinite nerve membrane. The FitzHugh–Nagumo equations consist of a nonlinear reaction–diffusion equation coupled to an

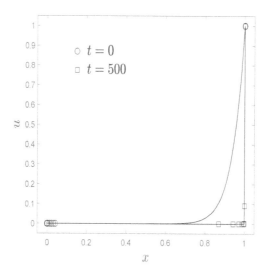

FIGURE 3.4

The boundary layer solution and nodes positions for dimensionless time $t = 0$ and $t = 500$. Adapted with permission from Coimbra et al. [16]. ©2001, Elsevier.

TABLE 3.3

MFEM parameters. A boundary layer problem.

Number of finite elements	8
Initial mesh	concentrated on the left
Approximation degree	4
Degree of quadrature formula	11
Penalty constants	$10^{-5}, 0, 10^{-3}, 10^{-2}, 10^{-5}, 10^{-5}$
Absolute and relative tolerances	
for ODE solver	$10^{-5}, 10^{-5}$

TABLE 3.4

Integration information. A boundary layer problem.

NST	NRE	NJE	CPU time(s)
1589	11756	258	0.25

ordinary differential equation and are a simplified version of the Hodgkin–Huxley equations proposed in 1952 by Hodgkin and Huxley [17]. A real nerve has a finite extent, so we consider the spatial variable on the interval $x \in [0, 120]$ and $t > 0$. The normalized system to be studied is that considered in [12, 18].

$$\frac{\partial v}{\partial t} = b(u - cv) \qquad (3.9)$$

$$\frac{\partial u}{\partial t} = d\frac{\partial^2 u}{\partial x^2} + u(u - a)(1 - u) - v. \qquad (3.10)$$

with $0 \leq x \leq 120$ and $t > 0$. Here d is the diffusion coefficient and a, b, c are positive constants. The variable u is an electrochemical potential and measures the potential across the cell membrane and v is a recovery variable that enables the system to return to its rest state. Parameter b represents the reciprocal of the time scale associated with the recovery of the nerve [12]. We study these equations with the Neumann boundary conditions,

$$\frac{\partial u}{\partial x}(0, t) = -\frac{I}{2}$$

$$\frac{\partial u}{\partial x}(120, t) = 0, \ t > 0. \qquad (3.11)$$

The stimulus I represents a constant current externally applied to the left end of the space domain. Initial conditions are

$$v(x, 0) = 0$$

$$u(x, 0) = 0 \qquad (3.12)$$

Notice that the Neumann boundary condition at the left endpoint of the spatial domain is not consistent with the initial conditions. For numerical simulations, we will adopt the particular values widely used in the literature from the work of Mitchell [19, 20],

$$d = 1 \quad a = 0.139 \quad b = 0.008 \quad c = 2.54 \quad I = 0.45. \qquad (3.13)$$

For the numerical simulation, the time interval considered is $0 \leq t \leq 280$. Two meshes are considered. $\mathcal{M}_1$ is the mesh associated with the recovery variable v and $\mathcal{M}_2$ is the mesh associated with the electrochemical potential, u. The mesh trajectories are presented in Figure 3.5.

The time interval considered is sufficiently large to capture the main features of the solution. Repetitive pulses are generated and propagate at regular intervals resulting in periodic waves as Figure 3.6 shows. Figure 3.6 presents the axial profiles for the electrochemical potential and for the recovery variable for $t = 40, 80, 120, 160, 200, 280$.

Concerning the application of MFEM code, Table 3.5 summarizes the MFEM parameters used for the numerical simulation, while Table 3.6 gives the integration information. For further details of the FitzHugh–Nagumo model, the reader is referred to [21, 22].

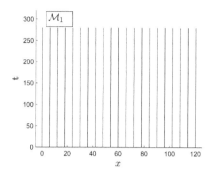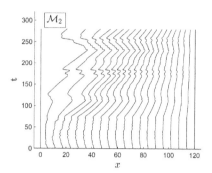

FIGURE 3.5
Electrical activity in a nerve membrane. Node movement for the mesh associated with recovery (left) and nodes movement for the mesh associated with potential variable (right).

3.5 Reactive–Diffusive System

The reaction–diffusion problem for a nonisothermal porous slab catalyst with half thickness ℓ, first-order exothermic reaction is studied. We consider the reaction–diffusion model in one dimension describing the concentration and temperature of a reactant in a single reaction $(A \rightarrow B)$. The dimensionless model [23] is

$$\frac{\partial c}{\partial t} = \frac{\partial^2 c}{\partial x^2} - \Phi^2\, c\, e^{\gamma-\gamma/T} \tag{3.14}$$

$$\frac{1}{Le}\frac{\partial T}{\partial t} = \frac{\partial^2 T}{\partial x^2} + \Phi^2 \beta\, c\, e^{\gamma-\gamma/T} \tag{3.15}$$

where x is the space variable normalized by ℓ, t is the time variable normalized by

$$\tau_d = \frac{\ell^2}{D_e}$$

c is the normalized concentration by the fluid concentration around the particle, c_0, and T is the normalized temperature by the initial temperature T_0. Moreover,

$\gamma = E/(RT_0)$ is the Arrhenius number

$\Phi^2 = \dfrac{\ell^2 k(T_0)}{De}$ is the square of Thiele modulus

$\beta = \dfrac{(-\Delta H)c_o D_e}{\lambda_e T_0}$, β is the Prater thermicity factor

$Le = \dfrac{\frac{\lambda_e}{\rho c_p}}{D_e}$ is the Lewis number

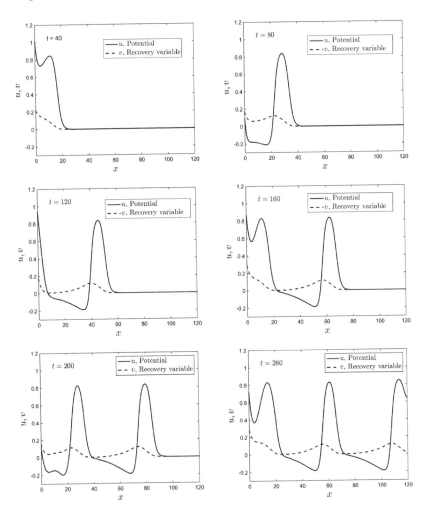

FIGURE 3.6
Electrical activity in a nerve membrane. Axial profiles for electrochemical po-
tential (solid) and for recovery variable (dash) for $t = 40, 80, 120, 160, 200, 280$.
Adapted with permission from Coimbra et al. [18]. ©2004, Elsevier.

Let us consider a symmetric slab geometry and the following assumptions:

1. $Le = 1$
2. Dirichlet boundary condition at $x = 1$, $c(1, t) = T(1, t) = 1$
3. Symmetry condition in the center, $\dfrac{\partial c(0,t)}{\partial x} = \dfrac{\partial T(0,t)}{\partial x} = 0$
4. Initial condition, $c(x, 0) = T(x, 0) = 1$

TABLE 3.5

MFEM parameters. A problem from mathematical biology.

Number of finite elements	$\mathcal{M}_1$: 20
	$\mathcal{M}_2$: 20
Initial mesh	equally spaced
Approximation degree	$\mathcal{M}_1$: 4
	$\mathcal{M}_2$: 4
Degree of quadrature formula	11
Penalty constants	$10^{-3}, 0, 10^{-2}, 10^{-2}, 10^{-5}, 10^{-5}$
Absolute and relative tolerances for ODE solver	$10^{-3}, 10^{-3}$

TABLE 3.6

Integration information. A problem from mathematical biology.

NST	NRE	NJE	CPU time(s)
1264	79155	384	15.3

Under the assumptions made above, the model reduces to one single equation. This can be done as follows: multiply Equation 3.14 by β then add the two equations, Equation 3.15 and Equation 3.14. This leads to

$$\frac{\partial}{\partial t}\left(\beta c + T\right) = \frac{\partial^2}{\partial x^2}\left(\beta c + T\right) \qquad (3.16)$$

From the corresponding steady-state problem it follows that $\frac{\partial^2}{\partial x^2}\left(\beta c + T\right) = 0$. So there exist constants C_1 and C_2 such as

$$\beta c + T = C_1 x + C_2 \qquad (3.17)$$

Applying the initial condition at $x = 0$ it follows that $C_2 = 1 + \beta$. The symmetry condition implies $C_1 = 0$, resulting that $\beta c + T = 1 + \beta$, which holds for $0 \leq x \leq 1$ and $t \geq 0$. Thus, under the assumptions made above, the model reduces to

$$\frac{\partial T}{\partial t} = \frac{\partial^2 T}{\partial x^2} + \Phi^2\left(1 + \beta - T\right)e^{\gamma - \gamma/T} \qquad (3.18)$$

with boundary conditions

$$\frac{\partial T(0, t)}{\partial x} = 0$$
$$T(1, t) = 1$$

The initial condition is $T(x, 0) = 1$.

TABLE 3.7

MFEM parameters. Reaction–diffusion model.

Number of finite elements	5
Initial mesh	concentrated on the left
Approximation degree	6
Degree of quadrature formula	33
Penalty constants	$10^{-4}, 0, 10^{-4}, 10^{-3}, 10^{-5}, 10^{-5}$
Absolute and relative tolerances for ODE solver	$10^{-12}, 10^{-12}$

This combustion model is particularly challenging for large values of γ since two phases can be distinguished in the solution: the ignition phase and the propagation phase. For small values of time, T gradually increases from $x = 0$. Then the ignition occurs, and the temperature jumps from the unity to $1 + \beta$. This example has been used by several authors to test their methods [12, 16, 24]. For all simulations concerning this example the MFEM parameters used are kept constant (see Table 3.7). Also, the numerical simulation was performed until a sufficiently large time, t, in order to capture the steady state.

For the first numerical simulation, case (i), we consider the values from [16], $\beta = 1$ and $\Phi = 0.5$ and $\gamma = 20$.

Due to the extremely fast rate of ignition, the formation of the front takes place very rapidly. Thus, small variable time-steps are required in order to get accurate solutions, so we need small ODE tolerances. For both meshes the starting node positions are concentrated at $x = 0$,

$$X_{i,1} = 0, X_{i,2} = 0.00002, X_{i,3} = 0.0001, X_{i,4} = 0.001, X_{i,5} = 0.005, X_{i,6} = 1.$$

With these values of β, γ, and Φ parameters the steady state occurs around $t = 0.29$. Figure 3.7 presents the axial profiles for temperature for $t = 0.26, 0.262, 0.27, 0.28, 0.29$.

A second simulation, case (ii), was done keeping the values of model parameters $\Phi = 0.5$, $\gamma = 20$ and taking a smaller value of β parameter, $\beta = 0.8$. The MFEM parameters are the same (Table 3.7). For $\beta = 0.8$ steady state occurs around $t = 0.4$.

Considering $\gamma = 20$ and $\Phi = 0.5$, Figure 3.8 compares the temperature histories at $x = 0$ for two different values of β parameter, $\beta = 1$ and $\beta = 0.8$.

In a third simulation, case (iii), we consider $\gamma = 30$, and the parameters used in case (i), $\beta = 1$ and $\Phi = 0.5$. With a higher value of γ, axial profiles became steeper and we can test once more the performance of MFEM code. For the numerical simulation we take the MFEM parameters as in the previous case (see Table 3.7). Figure 3.9 shows the dimensionless temperature T versus x at times $t = 0, 0.14, 0.1535, 0.154, 0.156, 0.157, 0.158, 0.16$.

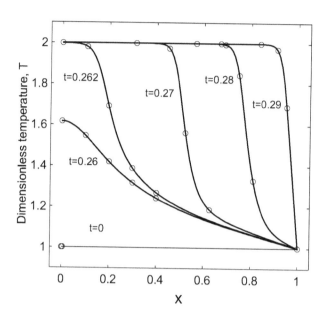

FIGURE 3.7

Reaction–diffusion model. Dimensionless temperature profiles at several time instants for $\beta = 1$, $\Phi = 0.5$ and $\gamma = 20$.

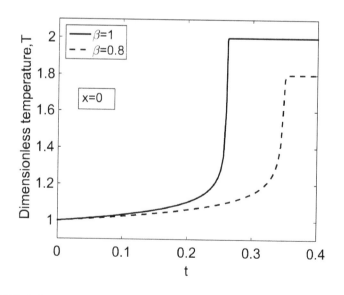

FIGURE 3.8

Reaction–diffusion model. Dimensionless temperature histories at $x = 0$ for $\gamma = 20$, $\Phi = 0.5$ and for $\beta = 1$ (solid) and $\beta = 0.8$.(dashed).

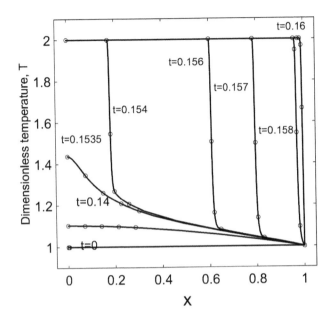

FIGURE 3.9

Reaction–diffusion model. Dimensionless temperature profiles at several time instants for $\beta = 1$, $\Phi = 0.5$, and $\gamma = 30$.

Figure 3.10 shows the nodes movement for $\gamma = 20$ and $\gamma = 30$. We can observe that the mesh produced by MFEM is following the dynamics of the problem. The instant at which ignition occurs can be detected by observing the nodes movement following the reaction front.

Table 3.8 presents the integration information for the three simulations presented. Simulations with other values for model parameters can be found in [16].

3.6 A Convection–Diffusion–Reaction Problem in a Catalytic Particle

In the next example we apply the MFEM to the classical engineering problem describing the normalized concentration, c, of a reactant in a catalytic particle [25]. Supposing that the reactant transport is by convection and diffusion, and a single first-order irreversible reaction occurs, the conservation of mass applied to the reactant leads to the linear convection–diffusion–reaction model

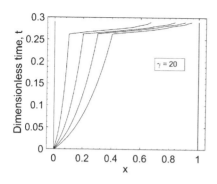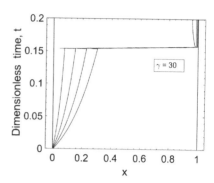

FIGURE 3.10
Reaction–diffusion model. Mesh trajectories, $\beta = 1$, $\Phi = 0.5$, and $\gamma = 20$ (left) and $\gamma = 30$ (right).

TABLE 3.8
Integration information. Reaction–diffusion model.

Simulations	NST	NRE	NJE	Time interval	CPU time(s)
case (i)	6807	27345	565	[0,0.29]	1.21
case (ii)	391	18987	391	[0, 0.4]	0.83
case (iii)	29335	157584	3628	[0,0.2]	6.82

for isothermal slab catalyst

$$\frac{\partial c}{\partial t} = \frac{\partial^2 c}{\partial x^2} - \lambda \frac{\partial c}{\partial x} - \Phi^2 c \qquad (3.19)$$

The dependent variable c is the normalized concentration by the concentration of the reactant in the fluid around the particle. The space variable, $0 \le x \le 2$ is the normalized space variable by the half thickness of the particle, ℓ. The time variable is normalized by the diffusion time constant,

$$\tau_d = \frac{\epsilon \ell^2}{D_e}, \qquad (3.20)$$

where ϵ is the particle porosity and D_e is the effective diffusivity. The intra-particle Peclet number, λ, is defined as

$$\lambda = \frac{v_0 \ell}{D_e} \qquad (3.21)$$

where v_0 is the intraparticle convective velocity. The Thiele modulus Φ is defined as

$$\Phi = \ell \sqrt{\frac{k}{D_e}} \qquad (3.22)$$

where k is the reaction constant. The problem is subjected to the initial condition,

$$c(x,0) = 0, \ 0 \leq x \leq 2, \tag{3.23}$$

and Dirichlet boundary conditions,

$$\begin{aligned} c(0,t) &= 1 \\ c(2,t) &= 1, \ t > 0. \end{aligned} \tag{3.24}$$

For the numerical simulations, we consider the values from [18, 26, 27]. Integration is carried out until the steady state is reached. MFEM parameters adopted are the same (see Table 3.9) for all simulations. A first simulation is performed with the purpose of illustrating the absence of convection taking $\lambda = 0$ and $\Phi = 1$, case(a). Figure 3.11 shows the concentration profiles taken at $t = 0, 0.01, 0.05, 0.2, 0.5$, and 1.8. The analytical stationary solution given by Equation 1.26 is the dashed line in Figure 3.11. For these values of parameters λ and Φ the steady state is attained around $t = 1.8$.

Taking into account convection and setting $\lambda = 10$, let us consider four distinct cases, cases (b) to (e), corresponding to the values of Thiele modulus

$$\Phi = 0.1; \ 1; \ 2; \ 3$$

Steady state is reached near $t = 0.5$ for the four simulations.

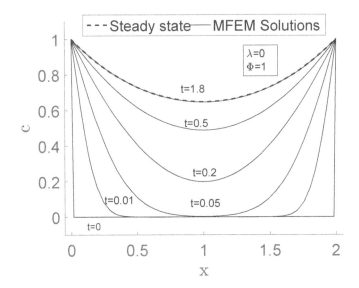

FIGURE 3.11
Diffusion-convection-reaction in a particle. Concentration profiles, $\lambda = 0$ and $\Phi = 1$ for different times from $t = 0$ to $t = 1.8$ and steady-state solution (dashed). Adapted with permission from Coimbra et al. [18]. ©2004, Elsevier.

Figure 3.12 and Figure 3.13 present the concentration profiles at dimensionless time, $t = 0, 0.001, 0.01, 0.05, 0.1, 0.2$, and 0.5 and show the effect of the Thiele modulus on the concentration profiles. Although it is more or less indistinguishable from the MFEM numerical solution at $t = 0.5$, the analytical stationary solution, given by Equation 1.34, is also represented (dashed line).

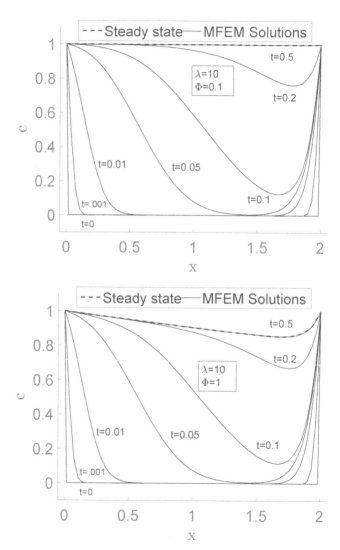

FIGURE 3.12

Diffusion-convection-reaction in a particle. Concentration profiles for $\lambda = 10$ and different values of Thiele modulus, $\Phi = 0.1$; 1, and steady-state solution (dashed). Adapted with permission from Coimbra et al. [18]. ©2004, Elsevier.

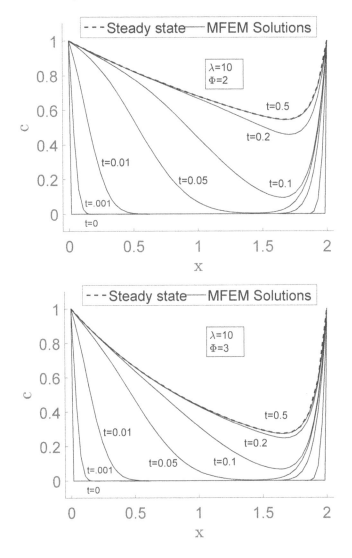

FIGURE 3.13
Diffusion-convection-reaction in a particle. Concentration profiles for $\lambda = 10$ and different values of Thiele modulus, $\Phi = 2$; 3, and steady-state solution (dashed). Adapted with permission from Coimbra et al. [18]. ©2004, Elsevier.

In the last numerical simulation of this problem, case (f), we consider the same value for λ parameter, $\lambda = 10$, and a higher value for Thiele modulus, $\Phi = 10$. Increasing the value of the Thiele modulus, the steady state is reached sooner, at $t = 0.07$, as Figure 3.14 illustrates. For these simulations, the perfect

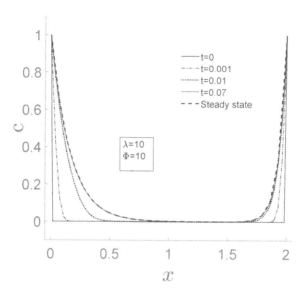

FIGURE 3.14
Convection–diffusion–reaction in a particle. Concentration profiles, $\lambda = 10$ and $\Phi = 10$ for $t = 0, 0.001, 0.01, 0.07$, and steady-state solution (dashed). Adapted with permission from Coimbra et al. [18]. ©2004, Elsevier.

agreement between the numerical solutions and the analytical solutions for the steady state demonstrates the reliability of the numerical simulations.

This example shows clearly the capacity of MFEM in reproducing the behavior of the convection–diffusion–reaction model. It is shown that MFEM is a robust numerical scheme, solving accurately with the same set of MFEM parameters (see Table 3.9) in a wide range of situations: the convection-free diffusion-reaction system, $\lambda = 0$, $\Phi = 1$, Figure 3.11; the convection domi-

TABLE 3.9
MFEM parameters. Convection–diffusion–reaction in a particle.

Number of finite elements	10
Initial mesh	equally spaced
Approximation degree	6
Degree of quadrature formula	33
Penalty constants	$10^{-4}, 0, 10^{-3}, 10^{-2}, 10^{-5}, 10^{-5}$
Absolute and relative tolerances for ODE solver	$10^{-12}, 10^{-12}$

TABLE 3.10

Integration information. Convection–diffusion–reaction in a particle.

Simulations	NST	NRE	NJE	Time interval	CPU time(s)
case (a)	4134	23783	286	$[0,1.8]$	1.83
case (b)	4848	27224	326	$[0,0.5]$	2.12
case (c)	4694	24722	293	$[0,0.5]$	1.94
case (d)	4563	22385	260	$[0,0.5]$	1.77
case (e)	4543	22564	264	$[0,0.5]$	1.77
case (f)	3758	19433	231	$[0,0.07]$	1.51

nated convection–diffusion–reaction problem, $\lambda = 10$, $\Phi = 0.1$, Figure 3.12; as well as the reaction dominated problem, $\lambda = 10$, $\Phi = 10$, Figure 3.14.

Table 3.10 gives the integration information for the six simulations presented.

3.7 Carbon Mask Adsorptive Reactor

In this example, we consider the model presented by Sereno [27], Rodrigues [28], and Soares et al., [29], a carbon mask adsorptive reactor for the elimination of cyanogen chloride, CNCl, from an air stream. The geometry of the model, with axial flow is illustrated in Figure 3.15.

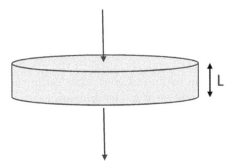

FIGURE 3.15

Carbon mask adsorptive reactor. Scheme of mask geometry.

This model is a modification of the original model developed by Friday [30] and includes the effect of axial dispersion. The authors concluded that axial dispersion does not alter the general behavior of the system. Nevertheless, the axial dispersion affects the breakthrough time of the mask considerably. Suppose a fixed bed of impregnated activated carbon where the impregnant is consumed in a reaction with the solute. Model equations are defined by taking into account the mass balance for solute in fluid phase and in solid phase and the mass balance for metal. First-order reactions are considered in each of the two reactants. Langmuir isotherm expresses the concentration of the adsorbed and gas phases in equilibrium. Here we consider dimensionless variables and parameters. The space variable is normalized by L, the bed length, so $0 \leq x \leq 1$. The time variable, $t \geq 0$ is the time reduced by the space time, $\tau = L/u_i$, where u_i is the interstitial velocity. The dimensionless dependent variables are the concentration in gas phase denoted by c, the concentration in the adsorbed phase, y, and the concentration of the metal impregnant, M. Concentrations are reduced by the inlet solute in the gas phase, C_{in}, by the solute concentration in the adsorbed phase in equilibrium with C_{in}, q_{in}, and the initial concentration of the metal, M_0, respectively. The normalized model equations are,

$$\frac{\partial y}{\partial t} = N_d (y - y^*) - Da \, y \, M \qquad (3.25)$$

$$\frac{\partial M}{\partial t} = -\alpha \, Da \, \chi \, y \, M \qquad (3.26)$$

$$\frac{\partial c}{\partial t} = \frac{1}{Pe} \frac{\partial^2 c}{\partial x^2} - \frac{\partial c}{\partial x} - N_f (c - c^*) \qquad (3.27)$$

and

$$N_f (c - c^*) = \xi N_d (y - y^*) \qquad (3.28)$$

The modified adsorption equilibrium isotherm is

$$y^* = \frac{K c^*}{1 + (K - 1)c^*} \qquad (3.29)$$

where c^* is the dimensionless concentration in the gas phase at the gas/solid interface and y^* is the dimensionless concentration in the adsorbed phase in equilibrium with c^*. The values of model dimensionless parameters are taken from the literature [28]:

Pe Peclet number, $Pe = 10^4$

N_f Number of film mass transfer units by film diffusion, $N_f = 74.5$

α Stoichiometric coefficient of the impregnant metal, $\alpha = 0.2$

Da Damköhler number, $Da = 1.575 \times 10^{-4}$, case (i) and $Da = 1.575 \times 10^{-5}$, case (ii)

χ ratio between the adsorbed phase concentration in equilibrium with the inlet solute concentration in the gas phase and the initial concentration of the impregnant metal, $\chi = 0.617$

N_d Number of intraparticle mass transfer units, $N_d = 2.5$

ξ bed capacity factor, $\xi = 4143$

K Modified constant of the adsorption isotherm, $K = 1.177$

The initial conditions for $t = 0$ and $0 < x < 1$ are

$$c = y = c^* = y^* = 0, \ M = 1 \tag{3.30}$$

and boundary conditions are defined by

$$
\begin{aligned}
\frac{\partial c}{\partial x}(0, t) &= (1 - c)\, Pe \\
\frac{\partial c}{\partial x}(1, t) &= 0, \ t > 0
\end{aligned} \tag{3.31}
$$

Figure 3.16 compares the evolution of the dimensionless concentration in the gas phase at the outlet, $x = 1$, as a function of the reduced time, for the two simulations, $Da = 1.575 \times 10^{-4}$ and $Da = 1.575 \times 10^{-5}$.

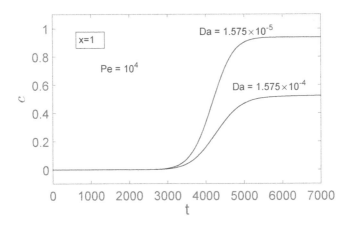

FIGURE 3.16
Concentration histories of a toxic gas in a carbon mask adsorptive reactor at $x = 1$ for $Da = 1.575 \times 10^{-4}$ and $Da = 1.575 \times 10^{-5}$.

Model equations are solved by using MFEM parameters described in Table 3.11. Mesh $\mathcal{M}_1$ denotes the mesh associated with the concentration in the adsorbed phase, y. Mesh $\mathcal{M}_2$ is associated with the concentration of metal impregnant, M, and finally mesh $\mathcal{M}_3$ is concerning the concentration of toxic species, c.

With respect to the adaptivity of the three meshes, Figure 3.17 illustrates the movements of the spatial nodes in the case of $Da = 1.575 \times 10^{-4}$, $\mathcal{M}_1$, $\mathcal{M}_2$ and $\mathcal{M}_3$ associated with the concentration in the adsorbed phase, the concentration of metal impregnant and the concentration of the toxic species, respectively.

Results for the dimensionless concentration profiles of toxic species in the fluid phase for both cases are shown in Figure 3.18. Notice that the

TABLE 3.11

MFEM parameters. Carbon mask adsorptive reactor.

Number of finite elements	$\mathcal{M}_1, \mathcal{M}_3$: 6
	$\mathcal{M}_2$: 2
Initial Mesh	concentrated on the left
Approximation degree	$\mathcal{M}_1, \mathcal{M}_3$: 6
	$\mathcal{M}_2$: 4
Degree of quadrature formula	11
Penalty constants	$10^{-5}, 0, 10^{-3}, 10^{-2}, 10^{-5}, 10^{-5}$
Absolute and relative tolerances	
for ODE solver	$10^{-5}, 10^{-5}$

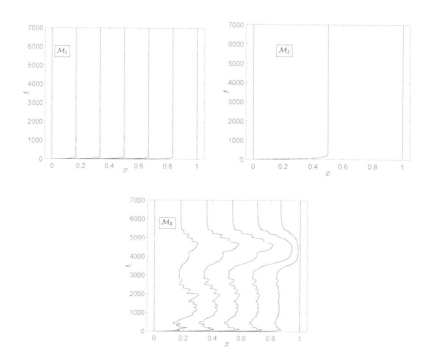

FIGURE 3.17

Carbon mask adsorptive reactor. Trajectories of nodes movements with time $(Da = 1.575 \times 10^{-4})$ for the mesh referring to the concentration in the adsorbed phase, $\mathcal{M}_1$, (top), the mesh concerning the concentration of metal impregnant, $\mathcal{M}_2$, (middle), and the mesh related to the concentration of the toxic species, $\mathcal{M}_3$(bottom).

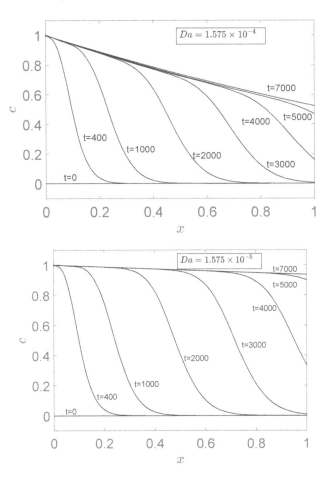

FIGURE 3.18
Concentration profiles of a toxic gas in a carbon mask adsorptive reactor,
$Da = 1.575 \times 10^{-4}$ (top) and $Da = 1.575 \times 10^{-5}$ (bottom).

concentration, c, of toxic species in the fluid phase is normalized by the inlet
solute concentration, c_{in} in the fluid phase.

Finally, Table 3.12 presents the integration information for the numerical
simulations.

TABLE 3.12
Integration information. Carbon mask adsorptive reactor.

Simulations	NST	NRE	NJE	CPU time(s)
case (i)	3511	202598	2141	23.46
case (ii)	3663	207479	2186	24.19

3.8   A Nonisothermal Tubular Catalytic Reactor Problem

Fixed-bed reactors are largely used in industrial catalytic processes and the work developed by Quinta Ferreira et al. [26, 31, 32] is of major relevance in the development of mathematical models of reaction and transport in catalyst systems containing large pores. In this proposed example, we consider the pseudo-homogeneous model to simulate a nonisothermal tubular catalytic reactor with axial dispersion in the fluid concentration [26]. Let us assume that c represents the fluid phase concentration normalized by feed concentration, c_{feed}, and T is the fluid temperature scaled by the feed temperature, T_{feed}.

The space variable x, $0 \leq x \leq 1$, is scaled by the reactor length, L, and the time variable $t \geq 0$ is normalized by the space time, $\tau = L/u_i$, where u_i is the interstitial velocity at the reactor entrance. The model equations include a mass balance equation and an energy balance equation. The dimensionless equations are:

$$\frac{\partial c}{\partial t} = \frac{1}{Pe}\frac{\partial^2 c}{\partial x^2} - \frac{\partial c}{\partial x} - Da\, c\, T\, e^{\gamma - \gamma/T} \tag{3.32}$$

$$\frac{\partial T}{\partial t} = \frac{\tau}{\tau_{hl}}\left[-\frac{\partial T}{\partial x} + \beta Da\, c\, T\, e^{\gamma - \gamma/T} - N_w\,(T - T_w) \right] \tag{3.33}$$

where T_w is the wall temperature scaled by T_{feed}. In this dimensionless model Pe is the Peclet number, Da the Damköhler number, γ the Arrhenius number, β is the normalized adiabatic temperature rise, τ_{hl} the time constant for thermal wave propagation and N_w the number of transfer units for the heat transfer at the wall. These dimensionless group are defined by

$$Pe = \frac{u_i L}{D_{ax}}, \quad \text{and} \quad Da = \frac{k(T_{feed})}{\epsilon}\tau$$

where D_{ax} is the axial dispersion coefficient, $k(T_{feed})$ is the rate constant at temperature T_{feed}, and ϵ is the bed porosity.

$$\beta = \frac{(-\Delta H)c_{feed}}{\rho_f C_f T_{feed}}, \quad \text{and} \quad \gamma = \frac{E}{RT_{feed}}$$

where ρ_f is the density of the fluid, C_f is the heat capacity of the fluid. $(-\Delta H)$ is the heat of reaction, E is the activation energy and R the gas constant. The time constant for thermal wave propagation is

$$\tau_{hl} = \frac{\epsilon \rho_f C_f + (1 - \epsilon)\rho_S C_S}{\epsilon \rho_f C_f}$$

where ρ_S is the density of the fluid, C_S is the heat capacity of the fluid, and

R_0 is the reactor radius. Finally, the number of transfer units for the heat transfer at the wall is given by

$$N_w = \frac{2U\tau}{\epsilon R_0 \rho_f C_f}$$

where U is the global heat transfer coefficient at the reactor wall. Furthermore, the initial conditions are

$$\begin{aligned} c(x,0) &= 0 \\ T(x,0) &= 1, \, 0 \leq x \leq 1 \end{aligned} \tag{3.34}$$

and the boundary conditions have the form

$$\begin{aligned} \frac{\partial c}{\partial x}(0,t) &= Pe\,(c-1) \\ \frac{\partial c}{\partial x}(1,t) &= 0 \\ T(0,t) &= 1, \, t > 0. \end{aligned} \tag{3.35}$$

According to the work of Sereno et al. [33] and Coimbra et al. [16], the MFEM parameters are given in Table 3.13. Spatial mesh $\mathcal{M}_1$ is the mesh associated to concentration and $\mathcal{M}_2$ is the mesh associated to temperature. Initial nodes, for both meshes, are positioned at

$$\begin{aligned} & X_{i,1} = 0, X_{i,2} = 5. \times 10^{-5}, X_{i,3} = 1. \times 10^{-4}, \\ & X_{i,4} = 1.5 \times 10^{-4}, X_{i,5} = 2. \times 10^{-4}, X_{i,6} = 2.5 \times 10^{-4}, \\ & X_{i,7} = 3. \times 10^{-4}, X_{i,8} = 3.5 \times 10^{-4}, X_{i,9} = 1 \end{aligned} \tag{3.36}$$

TABLE 3.13
MFEM parameters. A nonisothermal tubular catalytic reactor.

Number of finite elements	$\mathcal{M}_1$: 8
	$\mathcal{M}_2$: 8
Initial mesh	concentrated at $x = 0$
Approximation degree	$\mathcal{M}_1$: 4
	$\mathcal{M}_2$: 4
Degree of quadrature formula	11
Penalty constants	$10^{-5}, 0, 10^{-3}, 10^{-2}, 10^{-5}, 10^{-5}$
Absolute and relative tolerances	
for ODE solver	$10^{-5}, 10^{-5}$

The time integration interval is $t \in [0, 1500]$. Model parameters for the numerical simulation are [34]

$$\frac{\tau}{\tau_{hl}} = 2.08 \times 10^{-4}, \, \gamma = 21.8, \, N_w = 33.7, \, Pe = 10^4, \, Da = \beta = 0.7$$

and $T_w = 1$. Figure 3.19 (top) presents the axial profiles for concentration and Figure 3.19 (bottom) shows the axial profiles for temperature. The node positions are also marked in both graphics.

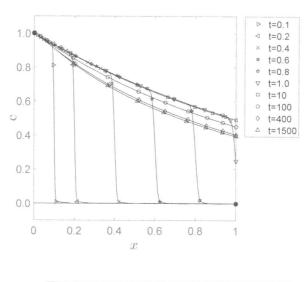

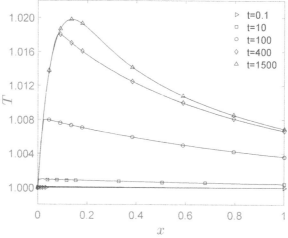

FIGURE 3.19

Nonisothermal tubular catalytic reactor. Concentration (top) and temperature (bottom) profiles as a function of the spatial variable. Adapted with permission from Coimbra et al. [16]. ©2001, Elsevier.

The MFEM gives an adequate representation of the steep concentration profiles occurring at short time and traveling to the right end of the bed. The concentration wave is traveling faster than the temperature wave. In the initial phase the reactor has a pseudo-isothermal behavior. Only much later, $t > 10$, the influence of the temperature is felt. During this second phase, concentration and temperature profiles follow the propagation of the thermal wave. Figure 3.20 shows the movement of mesh nodes with time for both meshes. Note that at short times nodes of $\mathcal{M}_1$ rapidly move to the right to capture the steep concentration profiles. Table 3.14 displays the integration information concerning the numerical simulation.

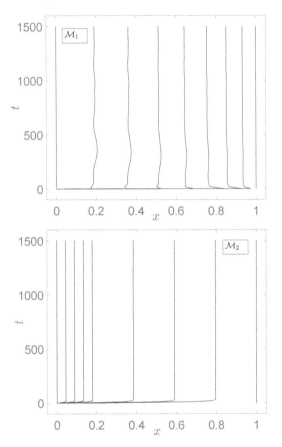

FIGURE 3.20
Nonisothermal tubular catalytic reactor. Mesh trajectories for concentration, $\mathcal{M}_1$ (top) and for temperature $\mathcal{M}_2$ (bottom).

Figure 3.21 presents the histories for concentration and temperature for $x = 0, 0.1, 0.2, 0.4, 0.6, 1$.

TABLE 3.14
Integration information. A
nonisothermal tubular catalytic reactor.

NST	NRE	NJE	CPU time(s)
1462	20513	167	2.66

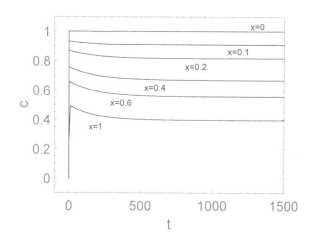

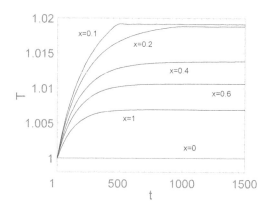

FIGURE 3.21
Nonisothermal tubular catalytic reactor. Concentration histories (top) and
temperature histories (bottom). Adapted with permission from Coimbra et
al. [16]. ©2001, Elsevier.

3.9 Separation of Optical Isomers by Chromatography

Chromatographic processes provide an important tool for the separation of multicomponent mixtures. This section presents a mathematical model describing the separation of isomers and the numerical solutions obtained by applying the MFEM code. This model considers a fixed bed chromatography column packed with adsorbent particles where two components are absorbed [35]. In the fluid phase, the two components are very diluted in an inert solvent. We also consider that the adsorption process is described by the equilibrium model, and a cylindrical geometry with a constant cross section and an isothermal process is assumed for the column.

To simplify the analysis, let us consider the independent dimensionless variable, x, the spatial variable normalized by the column length, L, and the normalized time variable, t, normalized by the space time $\tau = \epsilon L/u_s$, where ϵ is the bed porosity and u_s is the superficial velocity.

Let $c_i = c_i(x, t)$ represent the concentration of component i in the fluid phase and $q_i = q_i(x, t)$ denotes the concentration of component i in the solid phase. Moreover we consider the adsorption equilibrium isotherms, [36, 37]

$$q_i = k_i c_i + \frac{q_m b_i c_i}{1 + b_1 c_1 + b_2 c_2} \tag{3.37}$$

with $k_1 = k_2 = 1.57$, $q_m = 5.8$g/l, $b_1 = 0.045$ l/g and $b_2 = 0.127$ l/g. The governing equations under the above assumptions are,

$$\left(1 + \frac{1 - \epsilon}{\epsilon} \frac{\partial q_1}{\partial c_1}\right) \frac{\partial c_1}{\partial t} + \frac{1 - \epsilon}{\epsilon} \frac{\partial q_1}{\partial c_2} \frac{\partial c_2}{\partial t} = \frac{1}{Pe} \frac{\partial^2 c_1}{\partial x^2} - \frac{\partial c_1}{\partial x} \tag{3.38}$$

$$\frac{1 - \epsilon}{\epsilon} \frac{\partial q_2}{\partial c_1} \frac{\partial c_1}{\partial t} + \left(1 + \frac{1 - \epsilon}{\epsilon} \frac{\partial q_2}{\partial c_2}\right) \frac{\partial c_2}{\partial t} = \frac{1}{Pe} \frac{\partial^2 c_2}{\partial x^2} - \frac{\partial c_2}{\partial x} \tag{3.39}$$

where $Pe = \frac{L u_s}{\epsilon D_{ax}}$, is the Peclet number and D_{ax} is the axial dispersion coefficient. Moreover, let us consider two distinct sets of initial and boundary conditions corresponding to the problems

(i) Response of the fixed bed to a step in feed concentration.

(ii) Response of the fixed bed to a pulse in feed concentration.

For all numerical simulations, the MFEM parameters are those given in Table 3.15. Spatial mesh $\mathcal{M}_1$ is the mesh associated with the concentration of component 1 and $\mathcal{M}_2$ is the mesh associated with the concentration of component 2. Initial nodes, for both meshes, are concentrated at $x = 0$. Initially they are positioned at

$$X_{i,1} = 0, X_{i,2} = 1. \times 10^{-3}, X_{i,3} = 2. \times 10^{-3}, X_{i,4} = 3. \times 10^{-3},$$
$$X_{i,5} = 4. \times 10^{-3}, X_{i,6} = 5. \times 10^{-3}, X_{i,7} = 6. \times 10^{-3}, \tag{3.40}$$
$$X_{i,8} = 7. \times 10^{-3}, X_{i,9} = 8. \times 10^{-3}, X_{i,10} = 9. \times 10^{-3}, X_{i,11} = 1$$

TABLE 3.15

MFEM parameters. Separation of optical isomers by chromatography.

Number of finite elements	$\mathcal{M}_1:$ 10 $\mathcal{M}_2:$ 10
Initial mesh	concentrated at $x = 0$
Approximation degree	$\mathcal{M}_1:$ 11 $\mathcal{M}_2:$ 11
Degree of quadrature formula	33
Penalty constants	$10^{-7}, 0, 10^{-3}, 10^{-3}, 10^{-5}, 10^{-5}$
Absolute and relative tolerances **for ODE solver**	$10^{-3}, 10^{-5}$

Frontal chromatography of nonlinear systems. Response of the fixed bed to a step in feed concentration

Let us start with case (i) and assume that for $t < 0$ the column is fed with a pure solvent at flow rate Q. At $t = 0$ the column is fed with two components with concentrations c_{1_0} and c_{2_0}, diluted in an inert solvent. The corresponding initial conditions are

$$c_i(x, 0) = 0, \ i = 1, 2, \ 0 \leq x \leq 1 \qquad (3.41)$$

along with Robin boundary conditions,

$$\frac{\partial c_i}{\partial x}(0, t) = Pe(c_i - c_{i_0}) \qquad (3.42)$$

$$\frac{\partial c_i}{\partial x}(1, t) = 0 \qquad (3.43)$$

For the numerical simulations we use the data from [35], $\epsilon = 0.4$, $c_{1_0} = c_{2_0} = 5\text{g/l}$ and two different values for Peclet number, $Pe = 10^3$, case (ia), and $Pe = 10^4$, case (ib). For both cases the integration time interval considered is $t \in [0, 5]$.

In the absence of axial dispersion ($Pe = \infty$) model equations are hyperbolic and equilibrium theory predicts two shocks separated by a plateau [38]. Shocks occur at reduced time $t = 3.657$ and $t = 3.948$. The concentration at the intermediate plateau is $c_1 = 6.596\text{g/l}$ and $c_2 = 0\text{g/l}$. Figure 3.22 presents the breakthrough concentration curves at the bed outlet, $x = 1$, using $Pe = 10^3$ (top) and $Pe = 10^4$ (bottom) which are consistent with the theoretical findings.

Figure 3.23 gives the concentration profiles at $t = 0, 1, 2, 3, 3.5$, for both components, in the case of $Pe = 10^4$. It shows the formation of the fronts, moving to the right, resulting from two opposite contributions, a compressive front, and a dispersive front. The compressive front is rising from the favorable nature of the adsorption equilibrium isotherms while the dispersive front is

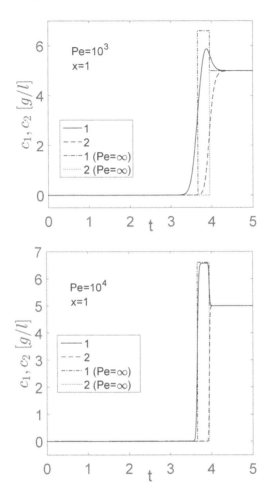

FIGURE 3.22
Frontal chromatography. Histories of concentration at $x = 1$ and solution from equilibrium theory, $Pe = 10^3$ (top) and $Pe = 10^4$ (bottom). Adapted with permission from Coimbra et al. [35]. ©2000, Elsevier.

originated by kinetic effects such as axial dispersion. Therefore, there is a time needed for the formation of the final fronts.

Node movement with time is illustrated in Figure 3.24. Initially, nodes are positioned at the inlet of the fixed bed and as integration starts they move to the right following the fronts.

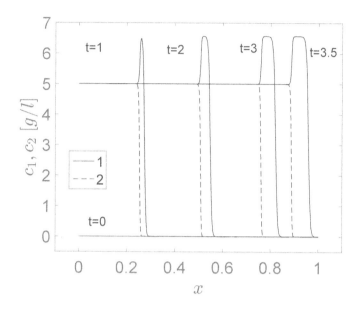

FIGURE 3.23
Frontal chromatography. Axial concentration profiles for component 1 and component 2 at dimensionless time $t = 0, 1, 2, 3, 3.5$ and $Pe = 10^4$. Adapted with permission from Coimbra et al. [35]. ©2000, Elsevier.

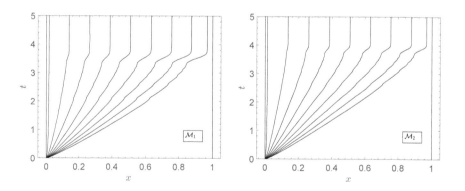

FIGURE 3.24
Frontal chromatography. Mesh trajectories for concentration of component 1, $\mathcal{M}_1$ (left) and concentration of component 2, $\mathcal{M}_2$(right), $Pe = 10^4$.

Elution chromatography of nonlinear systems. Response of the fixed bed to a pulse in feed concentration

For the second case, case (ii), we consider that the feed is done as in case (i) but only during a certain period of time, from $t = 0$ to t_{inj}. For $t > t_{inj}$ the column feed is changed to pure solvent. The feed flow rate is kept constant. Thus from $t > t_{inj}$ we must consider the concentration profiles at $t = t_{inj}$ as initial conditions and the boundary conditions change to,

$$\frac{\partial c_i}{\partial x}(0,t) = Pe\, c_i, t > t_{inj} \tag{3.44}$$

$$\frac{\partial c_i}{\partial x}(1,t) = 0, t > t_{inj} \tag{3.45}$$

As before, we use $\epsilon = 0.4$, $c_{1_0} = c_{2_0} = 5g/l$. Let us consider $Pe = 10^4$ and three different values for time injection: case (iia), $t_{inj} = 0.05$; case (iib), $t_{inj} = 0.5$, and case (iic), $t_{inj} = 1$. For all cases the integration time interval considered for the numerical simulation is $t \in [0,6]$.

Figure 3.25 compares concentration histories at $x = 1$ for these values of injection time.

Concentration profiles at $t = 0, 0.5, 2, 3$, for both components, are shown in Figure 3.26, for $t_{inj} = 0.5$.

Node movement with time is illustrated in Figure 3.27. Initially, nodes are positioned at the inlet of the fixed bed and, as integration starts, they move to the right following the front. Observing the node movement near $x = 0$ it is possible to identify the time instant at which the feed is changed to pure solvent. This shows the ability of MFEM to capture fine features in the solution.

Table 3.16 summarizes the integration information for the three simulations presented.

3.10 Pressurization of Adsorption Beds

Let us consider an example arising from pressure swing adsorption (PSA) process describing a model for the pressurization step.

The system considered here is shown in Figure 3.28. It consists of a packed bed of length L and cross-section area S. At the top and at the bottom there are two dead volumes with the same cross-section area and length L_1 and L_2, respectively.

The bed is packed with adsorbent particles of diameter d_p and the apparent density of the adsorbent is ρ_{ap}. The bed porosity is ε and ε_t is the total porosity. The model for pressurization is based on the following assumptions:

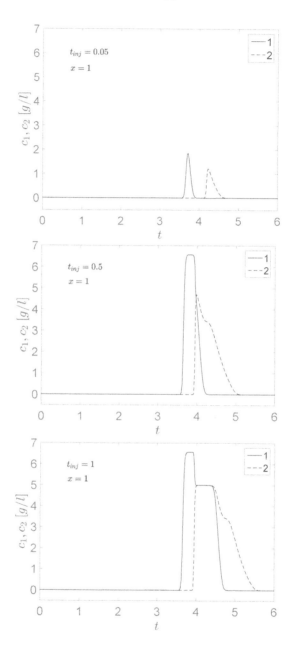

FIGURE 3.25
Elution chromatography. Histories of concentration at $x = 1$ for $t_{inj} = 0.05$ (top), $t_{inj} = 0.5$ (middle), and $t_{inj} = 0.5$ (bottom) and $Pe = 10^4$. Adapted with permission from Coimbra et al. [35]. ©2000, Elsevier.

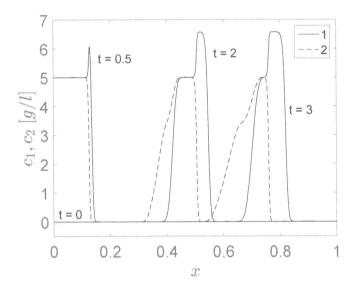

FIGURE 3.26
Elution chromatography. Axial concentration profiles for component 1 and component 2 at dimensionless time $t = 0, 0.5, 2, 3$ with $t_{inj} = 0.5$ and $Pe = 10^4$. Adapted with permission from Coimbra et al. [35]. ©2000, Elsevier.

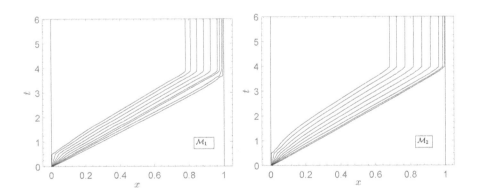

FIGURE 3.27
Elution chromatography. Mesh trajectories for concentration of component 1, $\mathcal{M}_1$ (left) and concentration of component 2, $\mathcal{M}_2$(right), $Pe = 10^4$.

TABLE 3.16

Integration information. Separation of optical isomers by chromatography.

Simulations	NST	NRE	NJE	Time interval	CPU time(s)
case (ia)	782	67633	275	[0,5]	82.93
case (ib)	1260	102884	418	[0,5]	125.19
case (iia)	1386	108675	441	[0,6]	131.18
case (iib)	2333	217171	884	[0,6]	266.32
case (iic)	2723	266702	1087	[0,6]	385.40

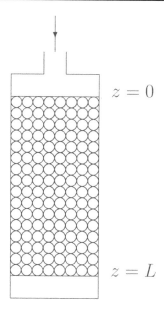

FIGURE 3.28

Sketch of pressurization of a packed-bed adsorber.

1. Isothermal operation.

2. Ideal gas behavior.

3. Subsonic flow and pressure drop versus flow condition relationships are instantaneously established locally in the bed.

Initially, the pressure of the system is P_l. At $t = 0$ the bed is fed with a binary mixture at a pressure P_h through a valve. Both P_l and P_h are expressed in atm. Here we consider the dimensionless model, so all variables are conveniently normalized. Independent variables are x and t. The spatial variable x is normalized by the column length, L, and the reduced time variable, t, is normalized by the space time $\tau = L/v_{ref}$. The reference velocity is chosen in such a way that

$$RT/(M_B v_{ref}^2) = 1 \qquad (3.46)$$

where M_B is the molecular mass of less strongly adsorbed component, R is the ideal gas constant, and T is the absolute temperature of the system. y_A represents the mole fraction of species A, the more strongly adsorbed component, and initially the mole fraction of species A is y_{A_0}.

The dependent variables, are the normalized partial pressure of species A, $p_A = p_A(x, t) = p y_A$, the normalized total pressure, $p = p_A + p_B = p(x, t)$ and the normalized mass flux, $G = \rho v$. Here ρ is the dimensionless fluid density and v is the dimensionless fluid velocity. Both p_A and p are normalized by $p_{ref} = P_h - P_l$. The dimensionless mass flux G is normalized by $G_{ref} = \rho_{ref} v_{ref} = \frac{p_{ref}}{v_{ref}}$. The dimensionless fluid density, ρ, is

$$\rho = p_A \left(\frac{M_A}{M_B} - 1 \right) + p \qquad (3.47)$$

where M_A is the molecular mass of more strongly adsorbed component.

Initial conditions for the model are,

$$p_A = y_{A_0} p_l, \quad p = p_l, \quad G = 0 \qquad (3.48)$$

where $p_l = P_l/p_{ref}$. At $t = 0$, the feed conditions at the inlet of top dead volume are:

$$p_{A_{feed}} = y_{A_{feed}} p_h, \quad p_{feed} = p_h \qquad (3.49)$$

where $p_h = P_h/p_{ref}$. Adsorption equilibrium is assumed to be represented by the Langmuir isotherm and for feed gases helium–methane mixtures we consider [39]

$$q_A = f(c_A) = \frac{5.94 \times 10^{-3} \times 4.62 c_A}{1 + 4.62 c_A} \qquad (3.50)$$

where c_A is the gas phase concentration of species A, methane, expressed in kmol/m^3 and q_A is the adsorbed phase concentration of A in kmol/kg.

The first model we consider, Model I, can be expressed by the dimensionless mass balance for species A equation,

$$\left(\epsilon_t + \rho_{ap}(1 - \epsilon) \frac{dq_A}{dc_A} \right) \frac{\partial p_A}{\partial t} = -\epsilon \frac{\partial}{\partial} \left(\frac{p_A G}{\rho} \right) + \frac{\epsilon}{Pe_{ref}} \frac{\partial}{\partial} \left(\frac{p^2}{\rho} \frac{\partial y_A}{\partial x} \right) \qquad (3.51)$$

the dimensionless total mass balance equation,

$$\left[\left(\epsilon_t + \rho_{ap}(1 - \epsilon) \frac{dq_A}{dc_A} \right) \frac{M_A}{M_B} \right] \frac{\partial p_A}{\partial t} + \epsilon_t \frac{\partial p}{\partial x} = \epsilon \frac{\partial G}{\partial x} \qquad (3.52)$$

and the dimensionless mechanical energy balance equation,

$$\frac{\partial G}{\partial t} = -\frac{\partial}{\partial x}\left(\frac{G^2}{\rho}\right) - \frac{\partial p}{\partial x} - \epsilon\frac{v_{ref}}{P_{BK}}\frac{\mu}{\mu_B}\frac{G}{\rho} + \text{sign}\left(\frac{\partial p}{\partial x}\right)1.75\frac{L}{d_p}\frac{G^2}{\rho}\frac{1-\epsilon}{\epsilon} \quad (3.53)$$

where μ_B is the viscosity of species B and $\text{sign}(\alpha) = \alpha/|\alpha|$, $\alpha \neq 0$.

We assume that μ, the viscosity of the binary mixture, is calculated by the Curtiss and Hirschfelder formula [40] and P_{BK} is the Blake–Kozeny parameter,

$$P_{BK} = \frac{d_p^2\epsilon^3}{150(1-\epsilon)}\frac{p_{ref}}{\mu_B L} \quad (3.54)$$

and Pe_{ref} is a reference Peclet number,

$$Pe_{ref} = \frac{Lv_{ref}}{D_{ax}} \quad (3.55)$$

where D_{ax} is the axial dispersion coefficient.

The boundary conditions are determined by the dead volumes. We define four extra dependent variables,

$$p_A|_{x=0^+},\ p|_{x=0^+},\ p_A|_{x=1^-},\ p|_{x=1^-} \quad (3.56)$$

denoting the dimensionless partial pressure of A and the total pressure at inlet of the bed, $x = 0^+$, and at the outlet of the bed, $x = 1^-$. At the boundaries, mass balances to species A and overall mass balance define four new ordinary differential equations,

$$\frac{L}{L_1}\frac{\partial p_A}{\partial t}\Big|_{x=0^+} = \frac{S_1}{S}\frac{y_{A_{feed}}}{y_{A_{feed}}\left(\frac{M_A}{M_B}-1\right)+1}G_1-$$

$$-\epsilon\frac{p_A|_{x=0^+}}{\rho|_{x=0^+}}G|_{x=0^+} + \frac{\epsilon}{Pe_{ref}}\frac{\omega|_{x=0^+}}{\rho|_{x=0^+}} \quad (3.57)$$

$$\frac{L}{L_1}\left(\frac{M_A}{M_B}-1\right)\frac{\partial p_A}{\partial t}\Big|_{x=0^+} + \frac{L}{L_1}\frac{\partial p}{\partial t}\Big|_{x=0^+} = \frac{S_1}{S}G_1 - \epsilon G|_{x=0^+} \quad (3.58)$$

$$\frac{L}{L_2}\frac{\partial p_A}{\partial t}\Big|_{x=1^-} = \epsilon\frac{p_A|_{x=1^-}}{\rho|_{x=1^-}}G|_{x=1^-} - \frac{\epsilon}{Pe_{ref}}\frac{\omega|_{x=1^-}}{\rho|_{x=1^-}} \quad (3.59)$$

$$\frac{L}{L_2}\left(\frac{M_A}{M_B}-1\right)\frac{\partial p_A}{\partial t}\Big|_{x=1^-} + \frac{L}{L_2}\frac{\partial p}{\partial t}\Big|_{x=1^-} = \epsilon G|_{x=1^-} \quad (3.60)$$

where ρ is defined by Equation 3.47 and

$$\omega = p_A\frac{\partial p}{\partial x} - p\frac{\partial p_A}{\partial x} \quad (3.61)$$

and S_1 is the cross-section area at the entry of the top dead volume. Assume that the feed stream crosses a valve with valve coefficient C_v. Thus the following valve equation is used to calculate the dimensionless mass flux at the entry of the top dead volume, G_1,

$$G_1 = \begin{cases} 1.410 \times 10^{-5}C, & p|_{x=0^+} \leq p_h \left(\frac{2}{\gamma+1}\right)^{\frac{\gamma}{\gamma-1}} \\ 1.663 \times 10^{-5}C\sqrt{1 - \left(\frac{p|_{x=0^+}}{p_h}\right)^2}, & p|_{x=0^+} > p_h \left(\frac{2}{\gamma+1}\right)^{\frac{\gamma}{\gamma-1}} \end{cases} \tag{3.62}$$

where $\gamma = c_p/c_v \approx 1.4$ is the specific heat ratio, and

$$C = p_h \frac{C_v}{S_1} \sqrt{y_{A_{feed}} \left(\frac{M_A}{M_B} - 1\right)} \tag{3.63}$$

The mathematical model involving Equation 3.51, Equation 3.52, and Equation 3.53, is a general model for pressurization of adsorption beds. After some manipulation model equations can be expressed in a suitable form to be discretized by MFEM resulting in an ODE system to be solved in time. The four equations describing the boundary conditions, Equation 3.57, Equation 3.58, Equation 3.59, and Equation 3.60, are added to the ODE system to produce the approximate solution.

Simplifications of Model I are considered to develop a second model, Model II. Let us assume that in the mechanical energy balance, Equation 3.53, the kinetic energy change is neglected and a steady-state mechanical energy balance is assumed. Thereafter Equation 3.53 is replaced by equation

$$0 = -\frac{\partial p}{\partial x} - \epsilon \frac{v_{ref}}{P_{BK}} \frac{\mu}{\mu_B} \frac{G}{\rho} + \text{sign}\left(\frac{\partial p}{\partial x}\right) 1.75 \frac{L}{d_p} \frac{G^2}{\rho} \frac{1-\epsilon}{\epsilon} \tag{3.64}$$

which is Ergun's equation used locally in the bed [41]. Model equations for Model II are now Equation 3.51, Equation 3.52, and Equation 3.64. Thus, the dimensionless mass flux can be calculated from Equation 3.64 as a function of p and p_A variables

$$G = \text{sign}\left(\frac{\partial p}{\partial x}\right) \left(a - \sqrt{a^2 + \Lambda \rho \left|\frac{\partial p}{\partial x}\right|}\right) \tag{3.65}$$

where

$$a = \frac{\epsilon^2 d_p v_{ref}}{3.5(1-\epsilon)LP_{BK}} \frac{\mu}{\mu_B} \tag{3.66}$$

and

$$\Lambda = \frac{\epsilon d_p}{1.75(1-\epsilon)L} \tag{3.67}$$

Numerical simulation for both models consider the following model parameters:

$$L = 0.914\text{m}; \ S_1/S = 0.01; \ V_1 = 9.2 \times 10^{-6}\text{m}^3; \ V_2 = 10.6 \times 10^{-6}\text{m}^3$$

$$\epsilon = 0.447; \ \epsilon_t = 0.776; \ \rho_{ap} = 850\text{kg m}^{-3}; \ d_p = 0.315 \times 10^{-3}\text{m}$$

$$T = 298\text{K}; \ P_h = 5\text{atm}; \ P_l = 1\text{atm}$$

$$M_A = 16.04\text{g/mol}; \ M_B = 4.003\text{g/mol}$$

$$\mu_A = 1.106 \times 10^{-3}\text{kg/mol/s}; \ \mu_B = 1.974 \times 10^{-3}\text{kg/mol/s} \tag{3.68}$$

and

$$Pe_{ref} = 4 \times 10^4, \ y_{A_0} = 1; \ y_{A_{feed}} = 0.5; \ C_v = 0.02 \tag{3.69}$$

The pressurization is stopped when pressure p at the closed bottom end of the column reaches an arbitrarily close value, p_{f_i}, to the feed pressure p_{feed}. Here we consider $p_{f_1} = 0.99p_h + 0.01p_l$ and $p_{f_2} = p_h - 0.01p_l$. Reduced time associated to p_{f_i} is referred to t_{f_i}. Numerical simulations were carried out for both models.

Table 3.17 shows MFEM parameters for Model I. Concerning the simplified model, Model II, the parameters are the same but only two meshes are needed, $\mathcal{M}_1$ associated with p_A and $\mathcal{M}_2$ associated with p. We have chosen the same initial node positions for each one of the space meshes,

$$X_{i,1} = 0, X_{i,2} = 5 \times 10^{-4}, X_{i,3} = 1 \times 10^{-3}, X_{i,4} = 2 \times 10^{-3},$$

$$X_{i,5} = 3 \times 10^{-3}, X_{i,6} = 4 \times 10^{-3}, X_{i,7} = 5 \times 10^{-3}, X_{i,8} = 1 \times 10^{-2}, X_{i,9} = 1$$

for $i = 1, 2, 3$.

TABLE 3.17
MFEM parameters. Pressurization of adsorption beds.

Number of finite elements	$\mathcal{M}_1:$ 8 $\mathcal{M}_2:$ 8 $\mathcal{M}_3:$ 8
Initial mesh	concentrated at $x = 0$
Approximation degree	$\mathcal{M}_1:$ 6 $\mathcal{M}_2:$ 6 $\mathcal{M}_3:$ 6
Degree of quadrature formula	11
Penalty constants	$10^{-4}, 0, 10^{-2}, 10^{-1}, 10^{-4}, 10^{-4}$
Absolute and relative tolerances for ODE solver	$10^{-5}, 10^{-5}$

The dimensional dependent variables are computed from the dimensionless values for plotted output of the model, and for comparison with reported literature [42, 43]. Note, however, that the MFEM calculations within both models are in terms of the dimensionless variables.

Figure 3.29 and Figure 3.30 show numerical simulation solutions obtained from Model I, case (i), and from Model II, case (ii), leading to similar results. Axial profiles of partial pressure of A and axial profiles of total pressure, at different times, are shown in Figure 3.29 for both models. Figure 3.30 shows axial profiles of mass flux and axial profiles of mole fraction of A for both models.

Also the values of t_{f_i} are very close: $t_{f_2} = 1465.57$ for case (i) and $t_{f_2} = 1464.98$ for case (ii). Another indicator used to compare Model I and Model II is the reduced time, t_c, at which the pressure $p|_{x=0^+}$ reaches the value $p_h \left(\frac{2}{\gamma+1} \right)^{\frac{\gamma}{\gamma-1}} \approx 0.53 p_h$. Once again we have similar values, $t_c = 94.84$ for case (i) and $t_c = 95.18$ for case (ii). Simulations of the models enable us to assess the validity of Model II in the pressurization step of a PSA cycle.

In Figure 3.31 histories of dimensional mass flux as a function of reduced time at different axial position, $x = 0; 0.4; 0.6; 1$ are represented.

TABLE 3.18
Stopping criteria. Pressurization of adsorption beds.

Simulations	Model	t_c	t_{f_1}	t_{f_2}
case (i)	I	94.84	1257	1465.57
case (ii)	II	95.18	1280	1464.98

Concerning the Peclet number used in both models it is a reference Peclet number. It relates to the Peclet number through the dimensional velocity, $Pe = \frac{Lvv_{ref}}{D_{ax}}$. Thus $Pe = vPe_{ref}$. Figure 3.32 presents the Peclet number as a function of reduced time at different reduced axial positions for case (ii), $Pe_{ref} = 4 \times 10^4$.

Although the solution computed for case (ii) with Model II is graphically identical to the solution computed with Model I, the integration information given in Table 3.19 shows that Model II yields a very satisfactory approximation with less computational effort.

The effect of the valve constant, C_v, is also studied. We consider model II and $C_v = 0.02$, (case (ii)) $C_v = 0.06$, (case (iii)) and $C_v = 0.1$ (case (iv)). The time interval considered is $[0, t_{f1}]$. By augmenting the value of C_v, the pressurization time decreases, and the pressure at $x = 0$ reaches p_h faster. Figure 3.33 shows total pressure as function of reduced time t at $x = 0$ (top) and at $x = 1$ (bottom) for different values of C_v.

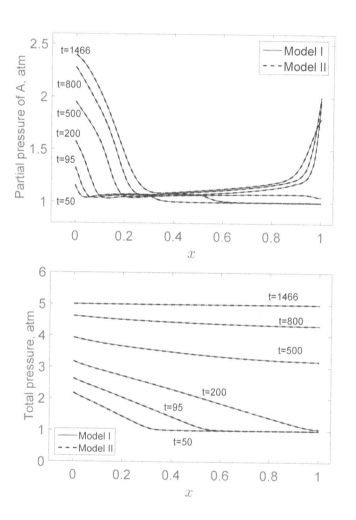

FIGURE 3.29
Pressurization of adsorption beds. Axial profiles of dimensional partial pressure of component A (top) and dimensional total pressure (bottom) for both models at several time instants.

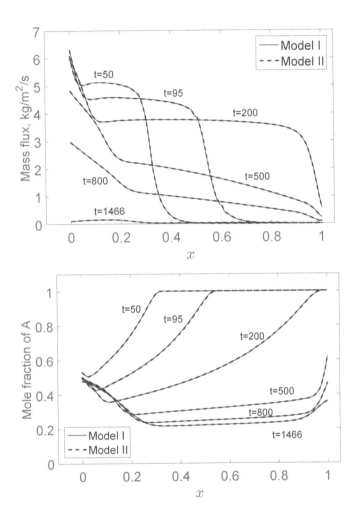

FIGURE 3.30
Pressurization of adsorption beds. Axial profiles of dimensional mass flux (top) and mole fraction of component A (bottom) for both models at several time instants.

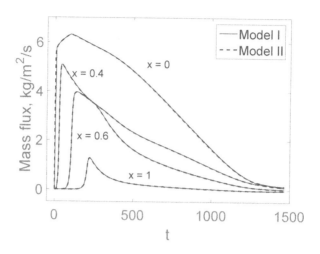

FIGURE 3.31
Pressurization of adsorption beds. Dimensional mass flux as a function of reduced time for both models at different reduced axial positions.

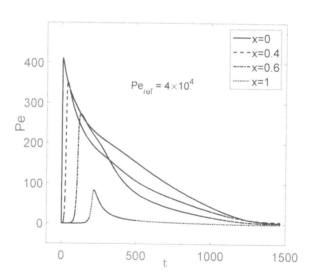

FIGURE 3.32
Pressurization of adsorption beds. Peclet number as a function of reduced time at different reduced axial positions, $Pe_{ref} = 4 \times 10^4$.

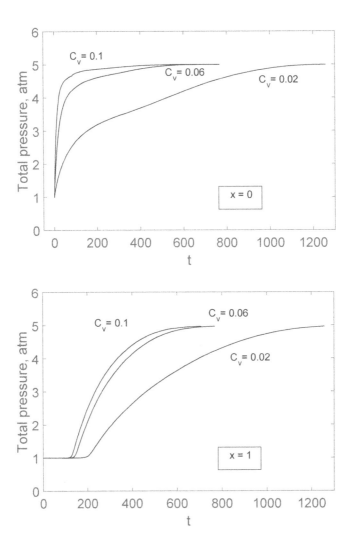

FIGURE 3.33
Pressurization of adsorption beds. Effect of C_v on the total pressure at $x = 0$ (top) and $x = 1$ (bottom) versus reduced time.

TABLE 3.19

Integration information. Pressurization of adsorption beds.

Simulations	Model	NST	NRE	NJE	Time interval	CPU time(s)
case (i)	I	5849	492530	2869	[0,1466]	134.2
case (ii)	II	720	27379	233	[0, 1465]	4.13
case (iii)	II	1593	70387	604	[0,767]	10.66
case (iv)	II	5683	454156	3948	[0,706]	67.72

3.11   Conclusions

The numerical algorithm based on the MFEM presented in Section 2.2 of Chapter 2 has been applied to the numerical simulation of several examples from science and chemical engineering. We have shown that MFEM code can effectively solve different transient problems with little or no user intervention. We propose a set of default penalty constants and ODE tolerances that can accommodate a large number of problems. The MFEM seems to be particularly well suited to solve problems where the combination of adaptive mesh and the robustness of the ordinary differential equations software can handle the sharp moving fronts to produce accurate solutions. The numerical simulations show the efficiency of MFEM code to solve convection–diffusion–reaction models.

Bibliography

[1] W. H. Press, B. P. Flannery, S.A. Teukolsky, and W. T. Vetterling. *Numerical Recipes in FORTRAN 77: The Art of Scientific Computing, 2nd Edition.* Cambridge University Press, 1992.

[2] J. Villadsen and M. L. Michelsen. *Solution of Differential Equation Models by Polynomial Approximation.* Prentice-Hall International Series in the Physical and Chemical Engineering Sciences. Englewood Cliffs, New Jersey: Prentice-Hall, Inc., 1978.

[3] A. C. Hindmarsh. LSODE and LSODIS, two new initial value ordinary differential equation solvers. *ACM-Signum Newslett*, 15:10–11, 1980.

[4] A. C. Hindmarsh. ODEPACK, a systematized collection of ode solvers. In R. S. Stepleman et al., editor, *Scientific Computing*, pages 55–64. Elsevier, 1983.

[5] MATLAB®. *Version 8.4.150421 (R2014b)*. The MathWorks Inc., Natick, Massachusetts, 2014.

[6] F. T. M. Nieuwstadt and J. A. Steketee. *Selected Papers of J. M. Burgers*. Springer Netherlands, 1995.

[7] C. Sereno, A. E. Rodrigues, and J. Villadsen. The moving finite element method with polynomial approximation of any degree. *Computers and Chemical Engineering*, 15(1):25–33, 1991.

[8] R. J. LeVeque. *Numerical Methods for Conservation Laws*. Birkhäuser Basel, 1992.

[9] Williem Hundsdorfer and Jan Verwer. *Numerical Solution of Time-Dependent Advection-Diffusion-Reaction Equations*. Springer, 2003.

[10] W. E. Schiesser and G. W. Griffiths. *A Compendium of Partial Differential Equation Models*. Cambridge University Press, 2009.

[11] Weizhang Huang and Robert D. Russell. *Adaptive Moving Mesh Methods*. Springer-Verlag New York, 2011.

[12] J. G. Verwer, J. G. Blom, and J. M. Sanz-Serna. An adaptive moving grid method for one-dimensional systems of partial differential equations. *Journal of Computational Physics*, 82(2):454–486, 1989.

[13] A. C. Mueller and G. F. Carey. Continuously deforming finite elements. *International Journal for Numerical Methods in Engineering*, 19:2099–2126, 1985.

[14] R. J. Gelinas, S. Doss, and K. Miller. The moving finite element method: Applications to general partial differential equations with multiple large gradients. *Journal of Computational Physics*, 40:202–249, 1981.

[15] N. N. Carlson and K. Miller. Design and application of a gradient-weighted moving finite element code i: In one dimension. *SIAM Journal on Scientific Computing*, 19(3):728–765, 1998.

[16] Maria do Carmo Coimbra, C. Sereno, and A. E. Rodrigues. Applications of a moving finite element method. *Chemical Engineering Journal*, 84(1):23–29, 2001.

[17] A. L. Hodgkin and A. F. Huxley. A quantitative description of membrane current and its application to conduction and excitation in nerve. *The Journal of Physiology*, 117(4):500–544, 1952.

[18] Maria do Carmo Coimbra, C. Sereno, and A. E. Rodrigues. Moving finite element method: Applications to science and engineering problems. *Computers and Chemical Engineering*, 28(5):597–603, 2004.

[19] A. R. Mitchell and V. S. Manoranjan. Finite element studies in reaction-diffusion. In *The Mathematics of Finite Elements and Applications IV*, pages 17–36. Academic Press, 1982.

[20] M. Bieterman and I. Babuska. An adaptive method of lines with error control for parabolic equations of the reaction–diffusion type. *Journal of Computational Physics*, 63(1):33–66, 1986.

[21] John Rinzel. Models in neurobiology. In R. H. Enns, B. L. Jones, R. M. Miura, and S. S. Rangnekar, editors, *Nonlinear Phenomena in Physics and Biology*, volume 75 of *NATO Advanced Study Institutes Series*, pages 345–367. Springer New York, 1981.

[22] J. D. Murray. Multi-species waves and practical applications. In J. D. Murray, editor, *Mathematical Biology*, volume 18 of *Interdisciplinary Applied Mathematics*, pages 1–70. Springer New York, 2003.

[23] A. K. Kapila. Reactive–diffusive system with arrhenius kinetics: Dynamics of ignition. *SIAM Journal on Applied Mathematics*, 39(1):21–36, 1980.

[24] S. Adjerid and J. E. Flaherty. A moving finite element method with error estimation and refinement for one-dimension time-dependent partial differential equations. *SIAM Journal on Numerical Analysis*, 23:778–796, 1986.

[25] A. Nir and L. Pismen. Simultaneous intraparticle forced convection, diffusion and reaction in a porous catalyst. *Chemical Engineering Science*, 32(1):35–41, 1977.

[26] R. M. Quinta Ferreira. *Contribuição para o estudo de reactores catalíticos em leito fixo: efeito da convecão em catalisadores de poros largos e casos de catalisadores bidispersos*. PhD thesis, Universidade do Porto, 1988.

[27] C. Sereno. *Método dos Elementos Finitos Móveis: Aplicações em Engenharia Química*. PhD thesis, Universidade do Porto, 1990.

[28] A. E. Rodrigues. Adsorptive separations and adsorptive reactors. In M.P.C. Weijnen and A.A.H. Drinkenburg, editors, *Precision Process Technology*, pages 327–344. Springer Netherlands, 1993.

[29] Rui C. Soares, Jose M. Loureiro, C. Sereno, and A. E. Rodrigues. Modeling and simulation of carbon mask adsorptive reactors. *Industrial and Engineering Chemistry Research*, 34(8):2762–2768, 1995.

[30] D. K. Friday. The breakthrough behavior of a light gas in a fixed-bed adsorptive reactor. *AIChE Symposium Series*, 84:89–93, 1988.

[31] R. M. Quinta Ferreira, A. C. Costa, and A. E. Rodrigues. Dynamic behavior of fixed-bed reactors with large-pore catalysts: A bidimensional heterogeneous diffusion/convection model. *Computers and Chemical Engineering*, 16(8):721–751, 1992. An International Journal of Computer Applications in Chemical Engineering.

[32] R. M. Quinta Ferreira, Paulo M. Simões, and A. E. Rodrigues. Simulation of tubular reactors packed with large-pore catalysts with spherical geometry. *Computers and Chemical Engineering*, 19, Supplement 1:351–356, 1995.

[33] C. Sereno, A. E. Rodrigues, and J. Villadsen. Solution of partial differential equations systems by the moving finite element method. *Computers and Chemical Engineering*, 16(6):583–592, 1992. An International Journal of Computer Applications in Chemical Engineering.

[34] R. M. Quinta Ferreira, Cristina A. Almeida-Costa, and A. E. Rodrigues. Effect of intraparticle convection on the transient behavior of fixed-bed reactors: Finite differences and collocation methods for solving unidimensional models. *Computers and Chemical Engineering*, 20(10):1201–1225, 1996.

[35] Maria do Carmo Coimbra, C. Sereno, and A. E. Rodrigues. Modelling multicomponent adsorption process by a moving finite element method. *Journal of Computational and Applied Mathematics*, 115(1-2):169–179, 2000.

[36] R-M Nicoud, G. Fuchs, P. Adam, M. Bailly, E. Küsters, F. D. Antia, R. Reuille, and E. Schmid. Preparative scale enantioseparation of a chiral epoxide: Comparison of liquid chromatography and simulated moving bed adsorption technology. *Chirality*, 5(4):267–271, 1993.

[37] A. E. Rodrigues, J. M. Loureiro, Z. P. Lu, and L. S. Pais. Modeling and operation of a simulated moving bed for the separation of optical isomers. In M.Douglas LeVan, editor, *Fundamentals of Adsorption*, volume 356 of *The Kluwer International Series in Engineering and Computer Science*, pages 765–772. Springer US, 1996.

[38] H. K. Rhee. Equilibrium theory of multicomponent chromatography. In A. E. Rodrigues and D. Tondeur, editors, *Percolation processes: Theory and Applications*. Sijthoff & Noordhoff, Alphen, 1980.

[39] H. C. Cheng and F. B. Hill. Separation of helium-methane mixtures by pressure swing adsorption. *AIChE Journal*, 31(1):95–102, 1985.

[40] R. Byron Bird, Warren E. Stewart, and Edwin N. Lightfoot. *Transport Phenomena*. New York: John Wiley, 2002.

[41] Z. P. Lu, J. M. Loureiro, A. E. Rodrigues, and M. D. LeVan. Pressurization and blowdown of adsorption beds - II. Effect of the momentum and equilibrium relations on isothermal operation. *Chemical Engineering Science*, 48(9):1699–1707, 1993.

[42] A. E. Rodrigues, J. M. Loureiro, and M. D. LeVan. Simulated pressurization of adsorption beds. *Gas Separation and Purification*, 5(2):115–124, 1991.

[43] C. Sereno and A. E. Rodrigues. Can steady-state momentum equations be used in modelling pressurization of adsorption beds? *Gas Separation and Purification*, 7(3):167–174, 1993.

4

Solving 2D Time-Dependent Problems

4.1 Introduction, Description, and Implementation

The moving finite element method described in Chapter 2 has been extensively tested for the numerical solution of convection–diffusion–reaction model in 1D spatial domains in Chapter 3. This chapter deals with the application of the moving finite element method to the two-dimensional spatial domains reported in Section 2.3.1. The implementation of the 2D MFEM algorithm uses a single spatial mesh to discretize the regular (rectangular) spatial domain even in the case of a system of PDEs. Also, a simplified set of penalty constants is used, and it is assumed that they are equal for all finite elements. The experience from the application of the 1D MFEM to different problems supports this option since the penalty constants are introduced only to prevent singularities of the matrix defining the system of ODEs obtained from the spatial discretization. Let us consider a finite element, Ω_i, of area $A(\Omega_j)$. The simplified version of Equation 2.40 is defined by the equations,

$$S_i = \frac{c_1}{A(\Omega_i) - c_3} \left(1 + \frac{c_3}{A(\Omega_i) - c_3} \right)^2 \tag{4.1}$$

$$\epsilon_i = \left(\frac{c_2}{A(\Omega_i) - c_3} + c_2 \right) \left(1 + \frac{c_3}{A(\Omega_i) - c_3} \right)^2 \tag{4.2}$$

where $c_i, i = 1, 2, 3$ are small positive constants supplied by the user. The constant c_3 represents the minimum allowed for an element area. The users of the 2D MFEM algorithm must also choose the degree for Gaussian quadrature formulae to be used in numerically solving integrals over 1D or 2D domains. The finite elements are triangles defined from a logically rectangular mesh by adding diagonals. Thus the same set of spatial nodes can produce two different meshes, depending on the orientation of the diagonals. By default, nodes are placed equally spaced, and the user may define the number of nodes in both axes. Figure 4.1 illustrates both cases for an equally spaced mesh with 72 triangles, i.e., 49 nodes over $\Omega = [0, 1] \times [0, 1]$.

Table 4.1 summarizes the 2D MFEM default parameters for the numerical simulations. As in Chapter 3, numerical simulations are based on the MFEM code in FORTRAN and the visualization of the results is carried out using the graphical capabilities of MATLAB®.

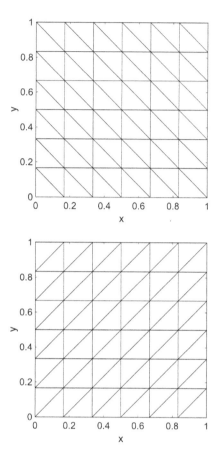

FIGURE 4.1
2D MFEM. Discretization of the spatial domain. Left-handed mesh (top) and right-handed mesh (bottom).

TABLE 4.1
2D MFEM default parameters.

Initial mesh	equally spaced
Approximation degree	2
Degree of 1D quadrature formula	11
Degree of 2D quadrature formula	7
Penalty constants	$10^{-6}, 10^{-4}, 10^{-6}$
ODE solver tolerances	$10^{-7}, 10^{-7}$

4.2 2D Burgers' Equation

As stated in Section 3.2, Burgers' equation is one of the most studied equations. For that reason, our first 2D example is the nonlinear Burgers' equation given by,

$$\frac{\partial u}{\partial t} = d\left(\frac{\partial^2 u}{\partial x^2} + \frac{\partial^2 u}{\partial y^2}\right) - u\left(\frac{\partial u}{\partial x} + \frac{\partial u}{\partial y}\right) \tag{4.3}$$

where $(x, y) \in [0, 1] \times [0, 1]$ and $t > 0$.

The MFEM techniques used by the 2D MFEM code are illustrated by considering the model defined by the scalar Equation 4.3 with initial and boundary conditions chosen such that the exact solution is

$$u(x, y, t) = \frac{1}{1 + e^{\frac{x+y-t}{2d}}} \tag{4.4}$$

This problem is often used to illustrate the adaptive solution techniques since for small values of the parameter, d, the partial differential equation is convection dominated, and the solution develops moving shock waves. It is used by Berzins et al. [1] as a text example.

Regarding the 2D MFEM simulations, we consider the case of $d = 0.025$, firstly. The solution is computed over the time interval $t \in [0, 2.5]$. The initial spatial mesh is obtained by adding right-handed diagonals to a rectangular grid. Figure 4.2 shows the MFEM solution as a function of (x, y) at time instants $t = 0$ and $t = 1$, as well as the spatial mesh where the solution is represented. Initially, a right-handed mesh is considered with 49 spatial nodes defining 72 finite elements. Quadratic approximations are used in each element.

We emphasize that when the time increases, the solution tends to $u = 1$. At time instant $t = 2$ we have $u(1, 1, 2) = 0.5$ while at $t = 2.5$ the solution is $u(x, y, 2) \approx 1, \forall (x, y) \in [0, 1] \times [0, 1]$. It can be observed that from $t = 2$ to $t = 2.5$ a redistribution of nodes. Actually, from $t = 2$ the node movement is done in a different way, typically to achieve a better position to represent accurately the plateau $u = 1$. This is what Figure 4.3 shows. On the top is displayed the mesh at time instant $t = 2$. The bottom figure shows the mesh at the final time instant, $t = 2.5$. It's important to note that at $t = 2$ nodes are concentrated near $(1, 1)$ where the solution has major variations. After that time instant, the redistribution of the nodes occurs.

To summarize the numerical simulation, Figure 4.4 displays the node movements and the MFEM solution for different instants of time, $t = 0.2$, $t = 0.3$, $t = 0.4$, $t = 0.5$, $t = 1.0$, $t = 1.4$, $t = 1.5$, and $t = 2.0$ with $d = 0.025$. It illustrates the capability of the MFEM to produce accurate solutions and to accurately track rapid spatial transitions without the need to adjust MFEM

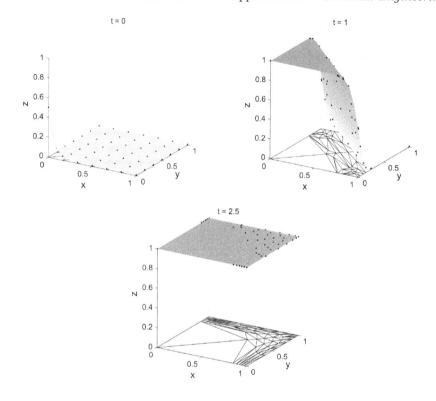

FIGURE 4.2
2D Burgers' equation. Spatial mesh and MFEM solution as a function of (x, y) at time instants $t = 0$ and $t = 1$ with $d = 0.025$.

parameters. Also, it must be noticed that starting with a right-handed spatial mesh the MFEM algorithm is able to reflect on the node movements the symmetry of the solution without the need of any extra condition.

More detailed numerical experiments, illustrating the connection between the node movement and the values of the penalty constant are presented by Coimbra et al. [3, 4].

A second numerical simulation takes a small value for diffusion parameter, $d = 0.005$. For the discretization of the spatial domain we choose a mesh with 81 spatial nodes, defining 128 finite elements and quadratic approximations. The problem is solved for $t \in [0, 1]$. Figure 4.5 compares the MFEM solution as a function of (x, y) at time instant $t = 1$, obtained with a right-handed mesh and with that achieved under the grid oriented to the left. The spatial mesh where the solution is obtained is also represented in Figure 4.5 for both cases. Table 4.2 outlines the integration information regarding the numerical simulations.

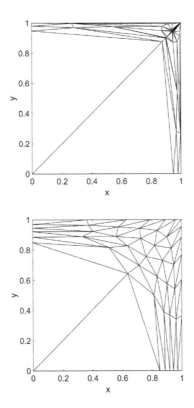

FIGURE 4.3
2D Burgers' equation. Spatial mesh at $t = 2$ (top) and $t = 2.5$ (bottom) with $d = 0.025$. Adapted with permission from Coimbra et al. [2]. ©2003, Elsevier.

4.3 2D Reaction–Diffusion Equation

Our second example is a more practical relevant problem described by Adjerid and Flaherty [5], Trompert and Verwer [6], Zegeling [7], and Coimbra et al. [3]. We revisit the reaction–diffusion model describing the concentration and temperature of a reactant in a single reaction, discussed in Chapter 3, Section 3.5 for one-dimensional space domain. Now the physical spatial domain has dimension 2 and by the assumptions made for one dimension, the dimensionless model is described by the single reaction–diffusion equation

$$\frac{\partial T}{\partial t} = \frac{\partial^2 T}{\partial x^2} + \frac{\partial^2 T}{\partial y^2} + \Phi^2\left(1 + \beta - T\right)e^{\gamma - \gamma/T} \qquad (4.5)$$

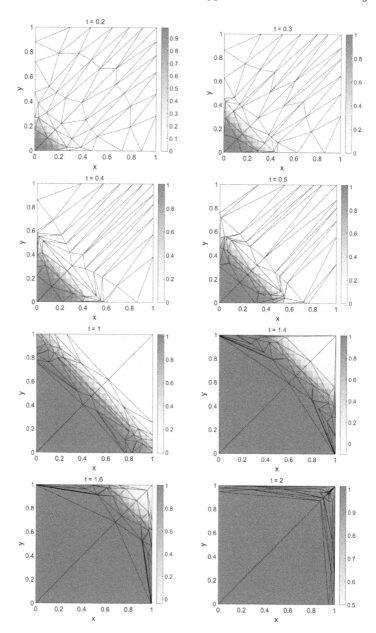

FIGURE 4.4

2D Burgers' equation. Node movements and MFEM solution for different instants of time, $t = 0.2$, $t = 0.3$, $t = 0.4$, $t = 0.5$, $t = 1.0$, $t = 1.4$, $t = 1.6$,
and $t = 2.0$, with $d = 0.025$. Adapted with permission from Coimbra et al.
[2]. ©2003, Elsevier.

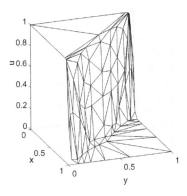

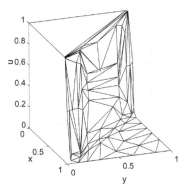

FIGURE 4.5
2D Burgers' equation. Spatial mesh and MFEM solution as a function of (x, y) at time instant $t = 1$, with $d = 0.005$ and a right-handed mesh (top) and a left-handed mesh (bottom).

TABLE 4.2
Integration information. Burgers' equation.

Simulations	Mesh	NST	NRE	NJE	CPU time(s)
$d = 0.025$	right-handed, 7×7	28690	903475	4474	3562
$d = 0.005$	right-handed, 9×9	7121	537344	1492	3923
$d = 0.005$	left-handed, 9×9	8898	714946	1986	4898

with $(x, y) \in [0, 1] \times [0, 1]$, $t > 0$ and subject to the following initial

$$T(x, y, 0) = 1, (x, y) \in [0, 1] \times [0, 1] \tag{4.6}$$

together with the boundary conditions

$$\frac{\partial T}{\partial x}\Big|_{(0,y,t)} = 0, \quad T(1, y, t) = 1, \quad 0 \le y \le 1, \, t > 0 \tag{4.7}$$

and

$$\left.\frac{\partial T}{\partial y}\right|_{(x,0,t)} = 0, \quad T(x,1,t) = 1, \quad 0 \le x \le 1, \ t > 0 \qquad (4.8)$$

The model parameters are the Prater thermicity factor, the Thiele modulus, and the Arrhenius number, β, Φ and γ, respectively, defined in Chapter 3, Section 3.5. The problem is solved for the parameters $\beta = 1$ and $\Phi = 0.5$ and two different values for γ parameter, $\gamma = 20$ and $\gamma = 30$.

Initially, the physical domain is at a constant temperature subject to Dirichlet and Neumann boundary conditions as Figure 4.6 shows. When the process starts the temperature gradually increases in the region around the origin. Then, suddenly, ignition occurs and the temperature jumps from a temperature near $T = 1$ to $T = 1 + \beta$ and a circular reaction front propagates toward the outer Dirichlet boundary where the temperature is kept constant. When the front reaches the boundary, a steady state results. The time instant at which the solution achieved the steady state depends on the model parameters. We choose a time interval large enough to guarantee that the numerical solution captures the behavior of the physical problem. Therefore, the MFEM solution is computed in the time interval $t \in [0, 0.4]$. The steady state is reached at time instant $t \approx 0.34$ for $\gamma = 20$ at time instant $t \approx 0.27$ for $\gamma = 30$.

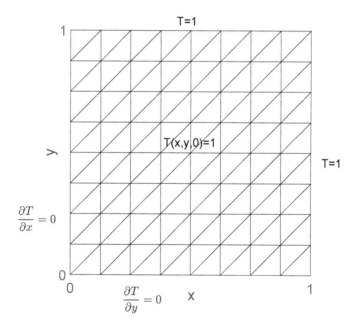

FIGURE 4.6
2D Reaction–diffusion equation. Discretization of the spatial domain with 128 finite elements.

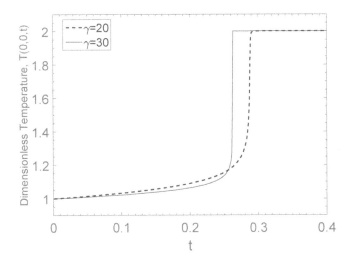

FIGURE 4.7
2D Reaction–diffusion equation. History of dimensionless temperature at $(0,0)$ as a function of the time variable.

Figure 4.6 gives the initial spatial mesh used for the numerical simulation. It is obtained by adding right-handed diagonals to a rectangular grid. The computation starts with a coarse mesh of 81 nodes defining 128 triangles.

Figure 4.7 compares the dimensionless temperature T as a function of the normalized time variable t at the origin of the spatial domain for $\gamma = 20$ and $\gamma = 30$. For $\gamma = 30$ a sharper solution can be observed. In fact, for $\gamma = 30$, a rapid variation in the temperature at $(0,0)$ from $t = 0.26$ to $t = 0.263$ is observed. The temperature jumps from $T = 1.2598$ to $T = 1.9997$. For $\gamma = 20$, the temperature jumps from $T = 1.254$ at time instant $t = 0.275$ to $T = 1.9996$ at time instant $t = 0.298$. The rapid variation in the solution illustrates the need for a robust numerical algorithm.

Usually, the simulation of flame propagation in combustion models requires a dense computational mesh in the thin flame region, and particularly in the boundary layer caused by the Dirichlet conditions. The 2D MFEM results of this example provide an indication of the advantages of MFEM in solving time-dependent problems with sharp moving fronts. Figure 4.8 shows the mesh evolution in time. The MFEM solution T is also plotted in a grayscale over the spatial mesh for different values of dimensionless time, $t = 0.29, 0.294, 0.298, 0.3, 0.32$, and $t = 0.4$. At time instant $t = 0.4$, the solution has already reached the steady state.

Figure 4.8 shows that the moving grid produced by the MFEM follows the dynamics of the problem. Moreover, the symmetry of the solution due to the initial and boundary conditions imposed is preserved in the node movement. Node movements and MFEM solution at different time instants, $t = 0.29, 0.294, 0.298, 0.3, 0.32$, and $t = 0.4$ are plotted in Figure 4.8.

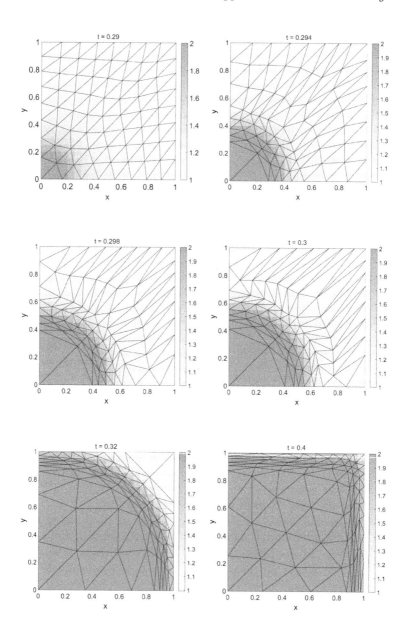

FIGURE 4.8

2D Reaction–diffusion equation. Node movements and MFEM solution at different time instants, $t = 0.29, 0.294, 0.298, 0.3, 0.32$, and $t = 0.4$.

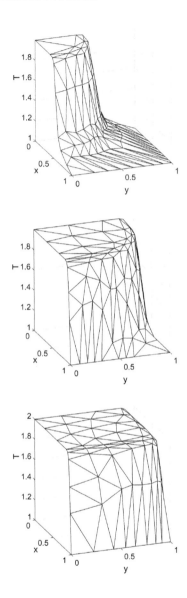

FIGURE 4.9
2D Reaction–diffusion equation. Dimensionless temperature as a function of the spatial variables at time instants $t = 0.3$, (top) $t = 0.32$ (middle), and $t = 0.34$ (bottom) with $\gamma = 20$.

Surface solution for time instants, $t = 0.3$, $t = 0.32$, and $t = 0.34$, with $\gamma = 20$ is presented in Figure 4.9. The effect of a higher value for γ parameter can be observed in Figure 4.10, where the dimensionless temperature as a

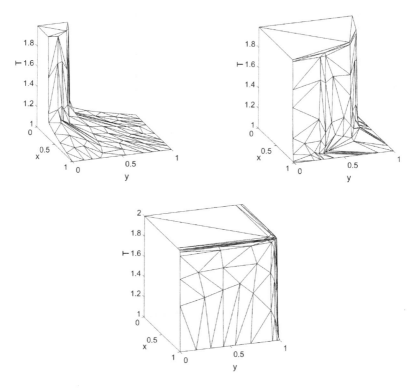

FIGURE 4.10
2D Reaction–diffusion equation. Dimensionless temperature as a function of the spatial variables at time instants $t = 0.263$, (top) $t = 0.265$ (middle), and (bottom) $t = 0.27$ with $\gamma = 30$.

function of the spatial variables at time instants $t = 0.263$, $t = 0.265$, and $t = 0.27$ with $\gamma = 30$ is displayed.

Table 4.3 outlines the integration information regarding the numerical simulations.

TABLE 4.3
Integration information. 2D reaction–diffusion equation.

Simulations	NST	NRE	NJE	CPU time(s)
$\gamma = 20$	6088	872925	2245	5088
$\gamma = 30$	14492	1331552	3410	7676

4.4 2D Burgers' System

We are now interested in the numerical simulation of the following two-dimensional Burgers' system

$$\frac{\partial u}{\partial t} = d \left(\frac{\partial^2 u}{\partial x^2} + \frac{\partial^2 u}{\partial y^2} \right) - u \frac{\partial u}{\partial x} - v \frac{\partial u}{\partial y} \tag{4.9}$$

$$\frac{\partial v}{\partial t} = d \left(\frac{\partial^2 v}{\partial x^2} + \frac{\partial^2 v}{\partial y^2} \right) - u \frac{\partial v}{\partial x} - v \frac{\partial v}{\partial y} \tag{4.10}$$

where $u(x, y, t)$ and $v(x, y, t)$ are dimensionless dependent variables and $(x, y) \in [0, 1] \times [0, 1]$ and $t > 0$ are the dimensionless independent variables. This couple of equations describes various kinds of phenomena. For example, Burgers [8] used it as a model to illustrate the theory of turbulence. Also, in the study of flow through a shock wave traveling in a viscous fluid this model is considered. In this case u and v represent the dimensionless velocity components.

Furthermore, initial and Dirichlet boundary conditions are chosen such that

$$u(x, y, t) = \frac{3}{4} - \frac{1}{4} \left(1 + e^{\frac{-4x + 4y - t}{32d}} \right)^{-1} \tag{4.11}$$

$$v(x, y, t) = \frac{3}{4} + \frac{1}{4} \left(1 + e^{\frac{-4x + 4y - t}{32d}} \right)^{-1} \tag{4.12}$$

define the exact solution [9]. Over the years, this example has often been used to test the accuracy of several numerical algorithms [10, 3, 11, 12, 13]. The solution represents two traveling waves with a sharp moving front, $y = x + \frac{t}{4}$. The speed of propagation is $\sqrt{2}/8$ and is perpendicular to the wave front [10].

The numerical computations over the time interval $[0, 3]$ with 2D MFEM were performed using a left-handed mesh with 100 spatial nodes defining 162 finite elements over $[0, 1] \times [0, 1]$ and cubic approximations in each element. The penalty constants are set to $c_2 = 10 \times -3, c_2 = 10 \times -2, c_3 = 10 \times -6$. Figure 4.11 gives the results for $d = 0.01$ showing the spatial mesh generated by the MFEM and both components of the solution, u and v as a function of the spatial variables (x, y) at different instants of time, $t = 0$, $t = 0.5$, $t = 1$, $t = 1.4$, $t = 2$, and $t = 3$. The CPU time for this run is 11,419 seconds. Similar numerical results can be achieved with different MFEM parameters. For example, by using a finite elements coarse mesh and quadratic approximations the CPU time falls down to 900 seconds.

A second run is accomplished to study the effect of d parameter. A small value is taken, $d = 0.005$. With a small value for the diffusion model parameter, d, the solutions exhibit sharper profiles than before. Nevertheless, keeping the MFEM parameters equal to those used in the first simulation, the results are very accurate as Figure 4.12 shows. The MFEM solutions at different instants of time as a function of the spatial variables, (x, y), for $d = 0.005$.

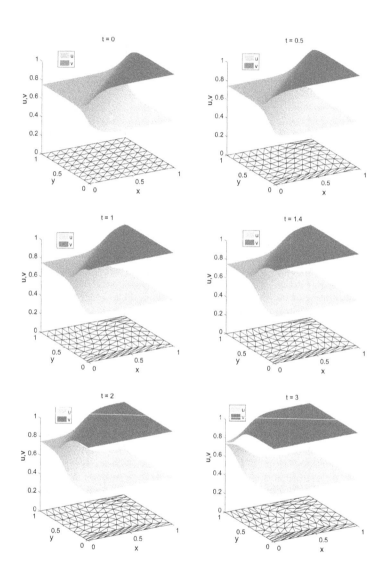

FIGURE 4.11
2D Burgers' system. Spatial mesh and MFEM solution as a function of (x, y) at time instants $t = 0$, $t = 0.5$, $t = 1$, $t = 1.4$, $t = 2$, and $t = 3$ with $d = 0.01$.

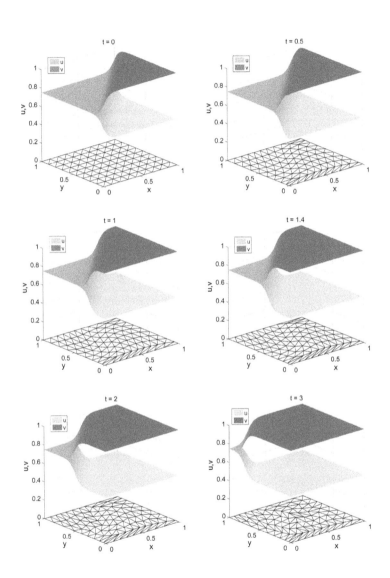

FIGURE 4.12
2D Burgers' system. Spatial mesh and MFEM solution as a function of (x, y) at time instants $t = 0$, $t = 0.5$, $t = 1$, $t = 1.4$, $t = 2$, and $t = 3$ with $d = 0.005$.

4.5   A Problem from Heat Transfer in a Fixed Bed

The study of heat transfer in packed beds is a topic of great interest to chemical engineers. We consider, in this example, the two-dimensional pseudo-homogeneous model proposed by Ferreira et al. [14, 15] for the transient heat transfer in a fixed bed packed with polymeric adsorbent particles. It is assumed that the transient temperature profiles depend on the axial and radial position in the column and, temperature differences between the fluid and solid phases are neglected. The axial variable z takes values from $z = 0$ (inlet) to $z = L$ (outlet). The column is assumed to be symmetrical so that the radial variable r ranges from the center of the column, $r = 0$, to the column wall $r = R_c$. The temperature, expressed in kelvins, is $T^*(z, r, \theta)$, where z, r are the spatial variables expressed in meters and θ is the time variable expressed in seconds.

For the pseudo-homogeneous model the energy balance gives

$$[\rho_f C_f \epsilon + \rho_S C_S (1 - \epsilon)] \frac{\partial T^*}{\partial \theta} =$$

$$K_{ae} \frac{\partial^2 T^*}{\partial z^2} - \epsilon u_i \rho_f C_f \frac{\partial T^*}{\partial z} + \frac{K_{re}}{r} \frac{\partial}{\partial r} \left(r \frac{\partial T^*}{\partial r} \right) \qquad (4.13)$$

The model parameters are:

- K_{ae}, the effective axial thermal conductivity, $Jm^{-1}s^{-1}K^{-1}$
- K_{re}, the effective radial thermal conductivity, $Jm^{-1}s^{-1}K^{-1}$
- u_i, the interstitial velocity, ms^{-1}
- ρ_f, the density of the fluid, kgm^{-3}
- ρ_S, the density of the solid, kgm^{-3}
- C_f, the heat capacity of the fluid, $Jkg^{-1}K^{-1}$
- C_S, the heat capacity of the solid, $Jkg^{-1}K^{-1}$
- ϵ is the bed porosity;

Furthermore, initial condition is

$$T^*(z, r, 0) = \begin{cases} T_0^* & \text{if } z > 0; \\ T_f^* & \text{if } z = 0. \end{cases} \qquad (4.14)$$

where T_0^* is the initial temperature, and T_f^* is the feed temperature. Boundary conditions are

$$T^*(0, r, 0) = T_F^* \qquad (4.15)$$

$$\frac{\partial T^*}{\partial z}(L, r, \theta) = 0 \qquad (4.16)$$

$$\frac{\partial T^*}{\partial r}(z, 0, \theta) = 0 \qquad (4.17)$$

and

$$-K_{re}\frac{\partial T^*}{\partial r}(z, R_c, \theta) = h_{we}\left(T^*(z, R_c, \theta) - T_a^*\right) \qquad (4.18)$$

where h_{we} is the overall heat transfer coefficient, $Jm^{-2}s^{-1}K^{-1}$, and T_a is the ambient temperature. Let us consider the dimensionless temperature,

$$T = \frac{T^* - T_a^*}{T_f^* - T_a^*} \qquad (4.19)$$

the dimensionless axial and radial coordinates,

$$x = \frac{z}{L} \qquad (4.20)$$

$$y = \frac{r}{R_c} \qquad (4.21)$$

and the dimensionless time variable

$$t = \frac{\theta u_i}{L} \qquad (4.22)$$

The dimensionless models are deduced by using the dimensionless groups, denoting the thermal axial Peclet number, the thermal radial Peclet number, the thermal Biot number, and the thermal capacity parameter, respectively.

$$Pe_{ha} = \frac{G_f C_f L}{K_{ae}} \qquad (4.23)$$

$$Pe_{hr} = \frac{G_f C_f L}{K_{re}} \qquad (4.24)$$

$$Bi_h = \frac{h_{we} R_c}{K_{re}} \qquad (4.25)$$

$$\xi_h = \frac{\rho_S C_S}{\rho_f C_f}\frac{1 - \epsilon}{\epsilon} \qquad (4.26)$$

where fluid $G_f = \epsilon u_i \rho_f$ is the fluid mass flux. The dimensionless equation reduces to

$$(1 + \xi_h)\frac{\partial T}{\partial t} = \frac{1}{Pe_{ha}}\frac{\partial^2 T}{\partial x^2} - \frac{\partial T}{\partial x} +$$
$$+ \frac{L}{R_c Pe_{hr}}\left(\frac{\partial^2 T}{\partial y^2} + \frac{1}{y}\frac{\partial T}{\partial y}\right) \qquad (4.27)$$

with the boundary conditions defined by

$$T(0, y, t) = 1 \qquad (4.28)$$

$$\frac{\partial T}{\partial x}(1, y, t) = 0 \qquad (4.29)$$

$$\frac{\partial T}{\partial y}(x, 0, t) = 0 \qquad (4.30)$$

At the column wall, the boundary condition is written as

$$\frac{\partial T}{\partial y}(x, 1, t) = -Bi_h T(x, 1, t) \tag{4.31}$$

The initial condition is

$$T(x, y, 0) = \begin{cases} T_0 & \text{if } 0 < x \leq 1; \\ 1 & \text{if } x = 0. \end{cases} \tag{4.32}$$

To overcome the discontinuity on the initial condition, the boundary condition $T(0, y, t) = 1$ is replaced by a slightly modified condition

$$T(0, y, t) = 1 + (T_0 - 1) e^{-10^3 t} \tag{4.33}$$

This abrupt initial profile requires a suitable spatial mesh (initial nodes concentrated near $x = 0$) for an accurate representation. Therefore a right-handed spatial mesh with 7×6 nodes is constructed by considering nodes with axial coordinates $x = 0, 0.025, 0.05, 0.075, 0.1, 0.2, 1$ and with radial coordinates $y = 0, 0.2, 0.4, 0.6, 0.8, 1$. Integration is performed for $0 \leq t \leq 4$.

Table 4.4 gives the parameters of the model to be used for the numerical simulation and Table 4.5 refers to the experimental conditions for the numerical simulations. Temperature is originally expressed in kelvins but data and temperature profiles are presented in the Celsius scale.

TABLE 4.4

Model parameters. Heat transfer in a fixed bed.

Cases	$L(m)$	$R_c(m)$	$T_a(°C)$	$T_F(°C)$	$T_0(°C)$
(i)	0.85	0.045	20	56	20
(ii)	0.85	0.045	24	56	20
(iii)	0.85	0.045	30	56	20

TABLE 4.5

Experimental conditions. Heat transfer in a fixed bed.

Pe_{ha}	Pe_{hr}	Bi_h	ξ_h
10^2; 10^3	500	8	1.3

Figure 4.13 compares the histories of the temperature as a function of the time variable for two different runs, $Pe_{ha} = 10^2$ and $Pe_{ha} = 10^3$. The histories of the temperature are taken at the average top, $x = 1$ and also at the middle of the column, $x = 0.5$, for the column center $y = 0$, and the column wall $y = 1$. The effect of a higher value for the thermal axial Peclet number is also present in Figure 4.14 where sharper axial profiles can be observed. Figure 4.14 gives the axial temperature profiles as a function of the axial coordinate

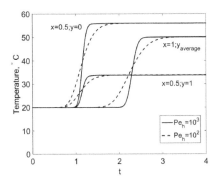

FIGURE 4.13
Heat transfer in a fixed bed. Histories of temperature with $Pe_{ha} = 10^2$ and $Pe_{ha} = 10^3$ for $T_a = 20°$C at the average top and at the middle of the column, center and wall.

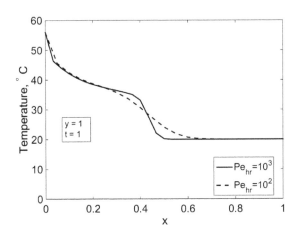

FIGURE 4.14
Heat transfer in a fixed bed. Axial temperature profiles with $Pe_{ha} = 10^2$ and $Pe_{ha} = 10^3$ for $T_a = 20°$C as a function of the axial coordinate x, at the wall of the column, $y = 1$, and at time instant $t = 1$.

x, at the wall of the column, $y = 1$, and at time instant $t = 1$, for $Pe_{ha} = 10^2$ and $Pe_{ha} = 10^3$ with $T_a = 20°$C.

The effect of the thermal axial Peclet number is shown in Figure 4.15, where the temperature profiles at time instant $t = 1$ as a function of the space variables, (x, y), are plotted.

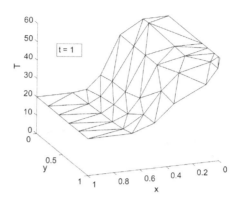

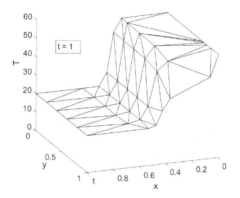

FIGURE 4.15

Heat transfer in a fixed bed. Temperature profiles at $t = 1$ as a function of (x, y) with $Pe_h = 10^2$ for $T_a = 20°C$.

Figure 4.16 shows the temperature profiles, as a function of the space variables, (x, y), at different instants of time for $Pe_{ha} = 10^2$. As expected, sharper profiles are obtained with a higher axial Peclet number $Pe_{ha} = 10^3$. Figure 4.17 gives the temperature profiles, as a function of the space variables, (x, y), at different instants of time, $t = 0.005$, $t = 1$, $t = 2.5$, $t = 2$, $t = 2.5$, and $t = 4$. Penalty constants are set to $c_1 = c_3 = 10^{-6}$ and $c_2 = 10^{-6}$ for all the numerical simulations and cubic approximations are used. The maximum CPU time is about 4,240 seconds. The reader will notice the similarity to how MFEM redistributes spatial nodes when the solution reaches the steady state. As in other problems, nodes follow the sharp moving front. When steady state is reached they adjust the position in order to achieve a desired fit. This can be seen in Figure 4.16 and Figure 4.17 from time instants $t = 2.5$ to $t = 4$.

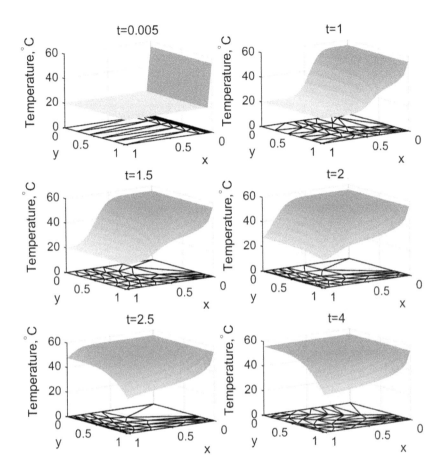

FIGURE 4.16
Heat transfer in a fixed bed. MFEM solution as a function of (x, y) at time instants $t = 0.005$, $t = 1$, $t = 2.5$, $t = 2$, $t = 2.5$, and $t = 4$ with $Pe_h = 10^2$ and $T_a = 20°C$. Adapted with permission from Coimbra et al. [4]. ©2004, Elsevier.

Let us consider the thermal axial Peclet number $Pe_{ha} = 10^3$ and two new simulations with $T_a = 24°C$, case (ii), and $T_a = 30°C$, case (iii), from Table 4.5. The effect of ambient temperature is studied, in particular when $T_a > T_0$. Figure 4.18 shows the histories of the temperature as a function of reduced time, taken at the middle of the column $x = 0.5$ and at the column wall

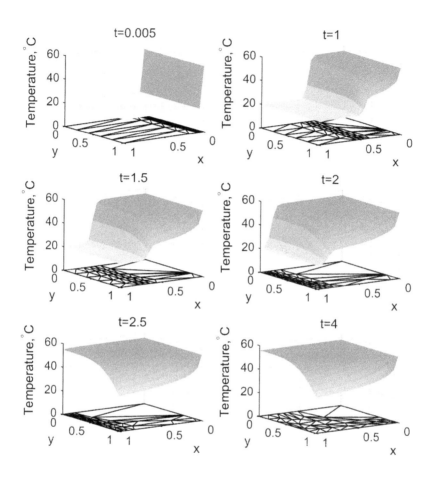

FIGURE 4.17
Heat transfer in a fixed bed. MFEM solution as a function of (x, y) at time instants $t = 0.005$, $t = 1$, $t = 1.5$, $t = 2$, $t = 2.5$, and $t = 4$ with $Pe_h = 10^3$ and $T_a = 20°C$.

$y = 1$. Finally, Figure 4.19 gives the temperature profiles, as a function of the space variables, (x, y), at time instant $t = 1$ and $t = 1.5$ with $T_a = 30°C$. The MFEM solutions are represented in the MFEM spatial mesh and node movements from $t = 1$ to $t = 1.5$ can be observed.

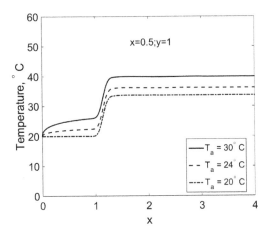

FIGURE 4.18
Heat transfer in a fixed bed. Histories of temperature at $(0.5, 1)$ (middle, wall) as a function of reduced time with $T_a = 30°C$, $T_a = 24°C$, $T_a = 20°C$, and $Pe_{ha} = 10^3$.

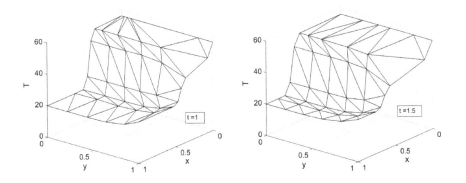

FIGURE 4.19
Heat transfer in a fixed bed. Temperature profiles at $t = 1$ and $t = 1.5$ as a function of (x, y) with $Pe_h = 10^3$ and $T_a = 30°C$.

4.6   Conclusions

We have presented the implementation of the MFEM for two-dimensional space domains and four problems have been solved with the MFEM algorithm. This 2D MFEM algorithm is obtained by incorporating the ideas of the 1D MFEM. The spatial domain is discretized by considering one spatial mesh and the spatial nodes move continuously to regions where they are most needed to represent the solution accuracy. At each time instant we require to find the nodes position of the mesh and the solution at the node positions by solving an ODE system. The numerical results are promising, demonstrating that 2D MFEM can significantly reduce the number of nodes needed to reach an accuracy solution of the convection–diffusion–reaction model in 2D spatial domains. We have illustrated how the moving finite element method can be used to solve nonlinear partial differential equations in two space dimensions. In particular, the accuracy of the numerical solutions indicates that the 2D MFEM is well suited for the solution of the two-dimensional models whose solution exhibits moving sharp front.

Bibliography

[1] M. Berzins, S. V. Pennington, P. R. Pratt, and J. M. Ware. Sprint2d software for convection dominated pdes. In Erlend Arge, AreMagnus Bruaset, and HansPetter Langtangen, editors, *Modern Software Tools for Scientific Computing*, pages 63–80. Birkhäuser Boston, 1997.

[2] Maria do Carmo Coimbra, C. Sereno, and A. E. Rodrigues. A moving finite element method for the solution of two-dimensional time-dependent models. *Applied Numerical Mathematics*, 44(4):449–469, 2003.

[3] Maria do Carmo Coimbra. *Método dos Elementos Finitos Móveis: Aplicação a sistemas de equações de derivadas parciais bidimensionais.* PhD thesis, Universidade do Porto, 2000.

[4] Maria do Carmo Coimbra, C. Sereno, and A. E. Rodrigues. Moving finite element method: Applications to science and engineering problems. *Computers and Chemical Engineering*, 28(5):597–603, 2004.

[5] Slimane Adjerid and Joseph E. Flaherty. A local refinement finite-element method for two-dimensional parabolic systems. *SIAM Journal on Scientific and Statistical Computing*, 9(5):792–811, 1988.

[6] R. A. Trompert and J. G. Verwer. A static-regridding method for two-dimensional parabolic partial differential equations. *Applied Numerical Mathematics*, 8(1):65–90, 1991.

[7] P. A. Zegeling. Moving-finite-element solution of time-dependent partial differential equations in two space dimensions. *International Journal of Computational Fluid Dynamics*, 1(2):135–159, 1993.

[8] J. M. Burgers. A mathematical model illustrating the theory of turbulence. In Richard Von Mises and Theodore Von Kámaán, editors, *Advances in Applied Mechanics*, volume 1, pages 171–199. Elsevier, 1948.

[9] Clive A. J. Fletcher. Generating exact solutions of the two-dimensional Burgers' equations. *International Journal for Numerical Methods in Fluids*, 3(3):213–216, 1983.

[10] J. G. Blom, R. A. Trompert, and J. G. Verwer. VLUGR2: A vectorizable adaptive grid solver for PDEs in 2D. *ACM Transactions on Mathematical Software*, 22(3):302–328, 1996.

[11] A.Refik Bahadir. A fully implicit finite-difference scheme for two-dimensional Burgers' equations. *Applied Mathematics and Computation*, 137(1):131–137, 2003.

[12] Hongqing Zhu, Huazhong Shu, and Meiyu Ding. Numerical solutions of two-dimensional Burgers' equations by discrete Adomian decomposition method. *Computers and Mathematics with Applications*, 60(3):840–848, 2010.

[13] Vineet K. Srivastava, Sarita Singh, and Mukesh K. Awasthi. Numerical solutions of coupled Burgers' equations by an implicit finite-difference scheme. *AIP Advances*, 3(8), 2013.

[14] L. M. Ferreira and A. E. Rodrigues. Adsorptive separation by thermal parametric pumping part I: Modeling and simulation. *Adsorption*, 1(3):213–231, 1995.

[15] L. M. Ferreira, J. A. M. Castro, and A. E. Rodrigues. An analytical and experimental study of heat transfer in fixed bed. *International Journal of Heat and Mass Transfer*, 45(5):951–961, 2002.

5

Solving Two Scales 1D+1d Time-Dependent Problems

5.1 Introduction, Description, and Implementation

This chapter is devoted to the application of the numerical scheme based on the moving finite elements method developed in Section 2.3.2 to solve time-dependent partial differential equations defined in a two-dimensional spatial domain with a specific 1D+1d structure, where it is possible to identify a macroscale and a microscale. For example, it is possible to find in some practical systems micro particles formed into macro-porous structures associated with transport scenarios, with the kinetics of those systems depending on the micro-pore and the macro-pore diffusional times.

Figure 5.1 shows a sketch of a two-scale 1D+1d problem. The bed length is the macroscale, so independent space variable z is defining the macro space domain. The radial position inside the particle defines the microscale.

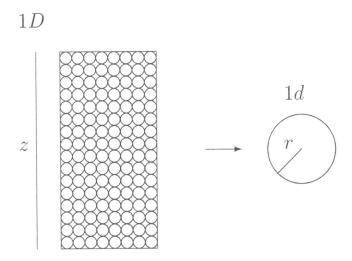

FIGURE 5.1
Two-scale 1D+1d time-dependent problem.

This new numerical procedure takes advantage of the well-known abilities of the MFEM and looks at the special structure of the spatial domain, to produce accurate solutions in less computational time compared with the 2D MFEM. In the present formulation the macro and the micro equations are discretized in space using 1D MFEM for each one of them and a crucial link is established between the two scales.

At each instant the MFEM computes the numerical solution of the PDE and also the spatial mesh where this solution is represented. In addition to the variants of the method proposed by Miller and coworkers, other alternatives have also been developed. Here we consider the use of higher-order Lagrange basis functions for both the trial and the test spaces and the use of a different spatial grid for each dependent variable [1, 2, 3, 4]. As shown in [5], the MFEM produces efficient and accurate numerical solutions for real problems in one or two space dimensions.

Looking more specifically to the 1D+1d problem, in first place the time-dependent macro equations are discretized in space using the 1D MFEM. Secondly, to each space node of the macro spatial mesh we associate a micro time-dependent problem, which is also discretized in space using the 1D MFEM. Then, we perform the links between the two scales [6] and finally the resulting system of ordinary differential equations is solved in order to get the numerical solution.

Examples that can be modeled as 1D+1d structures are found in fixed-bed processes. Usually other authors solve these fixed-bed process models without treating them as 1D+1d models. Raghavan and Ruthven [7] used the method of orthogonal collocation to predict the transient response of a fixed-bed adsorption column and the results were in good agreement with the analytical solutions. Won and Lee [8] proposed a numerical procedure to deal with steep transient concentration fronts using gradient-directed adaptive predictive collocation with a cubic spline interpolation function and far-side boundary conditions. Koocheksarayi et al. [9] used finite difference methods to study the transient behavior of sorption and intraparticle diffusion in fixed beds. Xu et al. [10] provided a review on mathematical models to describe the mass transfer and isotherm behavior of batch adsorption to predict the breakthrough curve. Afzal et al. [11] compared one-dimensional models with two-dimensional models of hydrogen fluoride adsorption by sodium fluoride, and concluded that the 1D model predicted well the adsorption behavior for lab-scale fixed beds.

So in a large class of two-dimensional problems it can be observed the presence of two different spatial scales, a macroscale and a microscale. With the aim of testing this new technique two examples were chosen. In the first example we portrait an absorbent or a catalyst constituted by spherical pellets placed into a packed column and contacted by a fluid. Mota et al. [12] modeled this as a transport problem involving two spatial scales. The overall dynamics of the packed bed determines the process and the packed bed is associated with the macroscale, while the microscale is associated with the pellets. The second example is the separation process of adsorption in a fixed bed, which is characterized by a concentration wave front, often a steep transient front that renders numerical computations difficult. We present the mathematical model for adsorption of phenol in a fixed bed of polymeric adsorbent under isothermal operation developed by Costa [13] and compare the results obtained.

Our numerical FORTRAN code is designed to solve 1D+1d problems where $u = u(z,t)$ is the dependent variable for the macro equation, while $u_p = u_p(r,z,t)$ is the dependent variable for the micro equations. The independent variables are the time variable, t, the normalized space variable in macro equation, z, and the normalized space variable in the micro equations, r.

The macro equation is given by the Equation 2.136

$$\frac{\partial u}{\partial t} = f\left(z,t,u,\frac{\partial u}{\partial z}\right)\frac{\partial^2 u}{\partial z^2} + h\left(z,t,u,\frac{\partial u}{\partial z}\right) + \vartheta\left(u_p(1,z,t)\right)$$

and it contains the linking term $\vartheta\left(u_p(1,z,t)\right)$ between the macro- and microscales. The micro equations must fulfill Equation 2.137, that is,

$$\frac{\partial u_p}{\partial t} = \bar{f}\left(r,t,u_p,\frac{\partial u_p}{\partial r}\right)\frac{\partial^2 u_p}{\partial r^2} + \bar{h}\left(r,t,u_p,\frac{\partial u_p}{\partial r}\right)$$

Furthermore Dirichlet boundary conditions or Robin boundary conditions are given. In particularly, for the micro equations at the external boundary, $r = 1$, the boundary condition is defined by Equation 2.138,

$$\frac{\partial u_p}{\partial r}(1,z,t) = \alpha\, u_p + \beta + \varpi\left(u(z,t)\right)$$

and it includes a term, $\varpi\left(u(z,t)\right)$ that links both scales. Therefore, the link between the two scales is done by the functions ϑ and ϖ. Finally we assume an initial condition to start the problem.

Similar to what was done earlier, we discretize the macro domain using a grid that moves with the solution of the macro problem. For each instant that grid causes the decomposition of the macro domain into finite elements. In each one of these we use Lagrange interpolation to find a polynomial approximation of the solution taking a collection of points in the finite element. Each point z_0 of the discretized macro domain is associated to a micro domain and a micro equation defined along it. We consider one micro problem for each grid point or interior point of a finite element or endpoints of macro domain. This is a dynamic choice to discretize an 1D+1d system. It is possible that z_0 for some time instant is associated to a micro problem and for other time instant it is not associated with any micro problem. Therefore, for each time instant t we have a collection of values $\{u_p(1,z_i,t)\}$ which are used to calculate $u_p(1,z,t)$ at a generic point z of the macro domain by using Lagrange interpolation. So, it is important that the collection of points $\{z_i\}$ in the macro domain associated to the micro problems is well distributed along the macro domain.

Preferentially, we opt for choosing an initial uniform grid in the macro domain and, it is desirable that its nodes have small movements so that the grid is almost uniform along time. We achieve this by taking a higher number of interior points in each finite element of the macro domain. There is a balance between the need of moving the nodes to steep gradient regions and the need of a good distribution along the macro domain. Another consequence of increasing the number of $1d$ problems associated with the smaller scale is the achievement of more accurate solutions. We define a concept of acceptable accurate solution in an iterative procedure, so there is a point from which there is no need to consider more $1d$ problems to obtain the desired solutions.

The next step is the discretization of the micro domains as in the MFEM. We choose a moving grid in each one of the micro domains. Note that there is no explicit link between the micro problems. Numerically, the link between both scales is one of the most important issues of our scheme. So we must implement correctly the terms ϑ and ϖ of the model equations.

If we are treating a micro equation associated with some z_0 in the macro domain and we want to implement the boundary condition at $r = 1$ we determine first $u(z_0, t)$ using Lagrange interpolation in the finite element of the macro domain that contains z_0 and then calculate the term $\varpi(y(z_0, t))$ that is added when the subroutine related to that boundary condition is called.

On the other hand, when we calculate the residual of the macro equation at some point z^* of the macro domain we have a collection of values $\{u_p(1, z_i, t)\}$ along the entire macro domain and using Lagrange interpolation we find an approximation of $u_p(1, z^*, t)$ which we use to determine $\vartheta(u_p(1, z^*, t))$. This last value is added in the subroutine of the code that calculates the residuals of the partial differential equations.

From the double discretization, macro domain and micro domains, results a large system of ordinary differential equations in time that is solved using the stiff integrator LSODIS [14]. In the vector of the time-dependent variables of the ordinary differential equations system we must know which block corresponds to which problem. The first block is for the macro problem and those following are for the micro problems. Concerning the MFEM penalty functions for $1D + 1d$ MFEM implementation the functions given by Equation 2.40 are used.

The $1D + 1d$ MFEM produced accurate results and showed to be a hopeful procedure for 1D+1d models that appear in many important examples in Chemical Engineering. Numerical results are presented in MATLAB® graphics.

5.2   Fixed-Bed Adsorption Model

Adsorbents and catalysts in industrial environments usually are constituted by spherical pellets placed into a packed column. The pellets are contacted by a fluid and two different diffusional resistances to mass transfer can be perceived. Mota et al. [12] modeled this as a transport problem involving two spatial scales. The overall dynamics of the packed bed determines the process and the packed bed is associated with the macroscale, while the microscale is associated with the pellets. The goal of numerical simulation is the study of the bulk concentration c, and the intraparticle concentration c_p of inert species in the packed bed, after a step change in the inlet concentration. These dimensionless variables are functions of the normalized space variable for the packed bed, z, the normalized space variable for the pellets, r, and the dimensionless time variable, t. For the definition of the dimensionless concentrations it is considered an initial concentration, C_0, and the inlet concentration, C_{in}. Dimensionless concentrations are defined by

$$c = \frac{C - C_0}{C_{in} - C_0} \quad \text{and} \quad c_p = \frac{C_p - C_0}{C_{in} - C_0} \tag{5.1}$$

The normalization of space variables is done with use of the radius of the pellets, R_p, and the length of the packed bed, L. The dimensionless time is normalized by $\tau = L/u_i$, where u_i is the interstitial velocity. The model contains four parameters: a parameter related to the axial dispersion, the Peclet number, Pe, a parameter related to the ratio of particle pore volume and interparticle (bed) pore volume, β, a parameter related to the diffusion in the pellets, α, and the Biot number Bi_m.

We define

$$Pe = \frac{\epsilon u_i L}{D_{ax}} \tag{5.2}$$

$$\beta = \frac{1 - \epsilon}{\epsilon} \epsilon_p \tag{5.3}$$

$$\alpha = \frac{D_{pe} L}{\epsilon_p u_i R_p^2} \tag{5.4}$$

$$Bi_m = \frac{k_f R_p}{D_{pe}} \tag{5.5}$$

where ϵ is the porosity of the packed bed, ϵ_p is the intraparticle porosity, D_{ax} is the axial dispersion coefficient, k_f is the film mass transfer, and D_{pe} is the effective pore diffusivity. Then the dimensionless equations that describe this $1D + 1d$ problem are:

$$\frac{\partial c}{\partial t}(z,t) = \frac{1}{Pe} \frac{\partial^2 c}{\partial z^2} - \frac{\partial c}{\partial z} - 3\beta\alpha Bi_m \left(c - c_p(1, z, t) \right) \tag{5.6}$$

$$\frac{\partial c_p}{\partial t}(r, z, t) = \alpha \left(\frac{\partial^2 c_p}{\partial r^2} + \frac{2}{r} \frac{\partial c_p}{\partial r} \right) \tag{5.7}$$

with $t > 0$ and $z, r \in [0, 1]$. The dimensionless concentrations of the inert species satisfy the following boundary conditions, for the packed bed ($z = 0$,

$z = 1$) and for the pellets ($r = 0$, $r = 1$):

$$\frac{\partial c}{\partial z}(0, t) = Pe\,(c - 1) \tag{5.8}$$

$$\frac{\partial c}{\partial z}(1, t) = 0 \tag{5.9}$$

$$\frac{\partial c_p}{\partial r}(0, z, t) = 0 \tag{5.10}$$

$$\frac{\partial c_p}{\partial r}(1, z, t) = Bi_m\,(c(z, t) - c_p(1, z, t)) \tag{5.11}$$

On the other hand, the initial concentrations are set to:

$$c(z, 0) = 0 \quad \text{and} \quad c_p(r, z, 0) = 0 \tag{5.12}$$

Within the formulation that we established for 1D+1d problems, we associate the packed-bed length to the macroscale and the radius of the spherical pellets to the microscale. Remember that the link between the two scales occurs at the external boundary of the micro domain. Thus, in the dimensionless equations defining the problem, the terms that link the macroscale to the microscale are:

$$\vartheta(z, t) = 3\beta\alpha Bi_m\, c_p(1, z, t) \tag{5.13}$$
$$\varpi(z, t) = Bi_m\, c(z, t) \tag{5.14}$$

Now, we have an interest in studying the effect of the parameter k related to mass transfer. For all the numerical simulations we assume the values for the model parameters given in Table 5.1. More detailed information can be found in [12].

TABLE 5.1
Model parameters. Fixed-bed
adsorption model.

Pe	β	α	Bi_m
40	0.5	0.2	0.2, 0.5, 0.8

Figure 5.2 shows the reduced time evolution of inlet dimensionless concentration; this is the dimensionless concentration at $z = 0$, for different values of Bi_m. For each case, plots of the dimensionless concentration at the center of the pellet, $r = 0$, and at the external boundary of the pellet, $r = 1$, are also presented.

To analyze the behavior of the model at the middle, $z = 0.5$, of the packed bed, Figure 5.3 gives the histories of dimensionless concentration as a function of reduced time at $z = 0.5$. The breakthrough curves are presented in Figure 5.4 showing the time evolution of the outlet dimensionless concentration, $z = 1$.

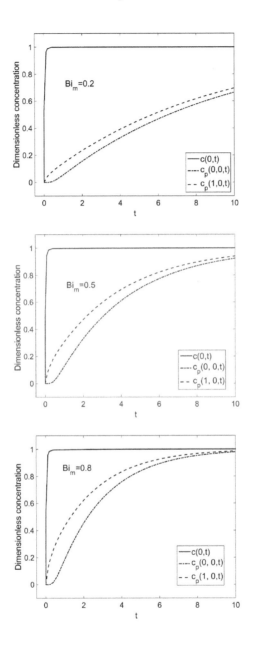

FIGURE 5.2

Fixed-bed adsorption model. Inert species dimensionless concentration as a function of dimensionless time, at $z = 0$, in the packed bed and at the center of the pellet, $r = 0$, and at the external surface of the pellet, $r = 1$, with $Bi_m = 0.2$ (top), $Bi_m = 0.5$ (middle) and $Bi_m = 0.8$ (bottom).

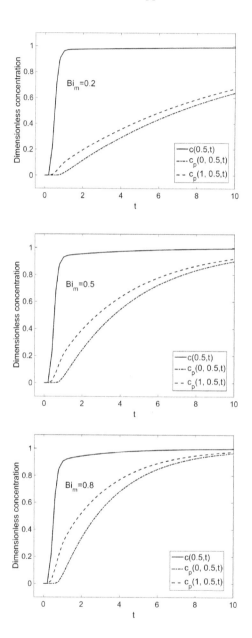

FIGURE 5.3
Fixed-bed adsorption model. Inert species dimensionless concentration as a function of dimensionless time, at $z = 0.5$, in the packed bed and at the center of the pellet, $r = 0$, and at the external surface of the pellet, $r = 1$, with $Bi_m = 0.2$ (top), $Bi_m = 0.5$ (middle) and $Bi_m = 0.8$ (bottom).

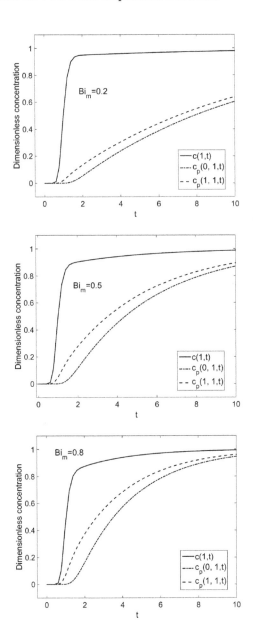

FIGURE 5.4

Fixed-bed adsorption model. Inert species dimensionless concentration as a function of dimensionless time, at $z = 1$, in the packed bed and at the center of the pellet, $r = 0$, and at the external surface of the pellet, $r = 1$, with $Bi_m = 0.2$ (top), $Bi_m = 0.5$ (middle) and $Bi_m = 0.8$ (bottom).

Concerning the numerical simulations with MFEM, we take a grid for the macro domain (packed bed) with 5 finite elements and 10 interior points per element. The micro domains (pellets) are decomposed in 4 finite elements with 4 interior points in each finite element. The number of micro problems equals the number of global nodes in the macro domain, so in this case we have 56 micro problems. The nodes of the initial grid in the macro domain are $z_1 = 0, z_2 = 0.2, z_3 = 0.4, z_4 = 0.6, z_5 = 0.8, z_6 = 1$. The nodes of the initial grid in the micro domains are positioned at $r_1 = 0, r_2 = 0.001, r_3 = 0.5, r_4 = 0.999, r_5 = 1$

As mentioned before, we chose a well-distributed initial grid for the macro domain, with several interior points in each finite element, to achieve an accurate approximation of the intraparticle concentration, c_p, at the boundary of the micro domain, $r = 1$. Remember that the grid in the macro domain is an adaptive grid, moving the nodes to regions with steep gradients in the concentration. It is not desirable to have a big gap between two consecutive nodes of the macro domain. However, similar solutions are obtained by choosing less interior points, for example, with only 4 interior points per finite element in the macro domain the results obtained are very similar. The integrals appearing in the equations of the ODE systems are computed by means of the Lobatto's quadrature with 11 interior quadrature points. For all simulations we use the default MFEM parameters,

$$c_1^{m,j} = c_5^{m,j} = c_6^{m,j} = 10^{-5}, c_2^{m,j} = 0, c_3^{m,j} = 10^{-3}, c_4^{m,j} = 10^{-2}$$

The absolute and relative tolerances of the ODE solver are set as $tol_1 = tol_2 = 10^{-5}$. It takes approximately $450s$ to compute each numerical simulation. If we chose tighter tolerances, the computational time has a notable increase. For example, if we set $tol_1 = tol_2 = 10^{-9}$, the CPU time increases to $1400s$.

By definition, the parameter Bi_m equals the product of the pellets radius and a ratio between the mass transfer coefficient at the external surface of the pellets and effective diffusivity inside the pellets. Therefore, increasing the parameter Bi_m implies a higher mass transfer through the film around the pellets. So for higher values of parameter Bi_m, the results show higher concentration of the inert species in the pellets and smaller concentration in the packed bed. The concentration profiles in the pellets and in the packed bed at the instant $t = 0.75$, for the values $Bi_m = 0.2$, $Bi_m = 0.5$ and $Bi_m = 0.8$ are presented in Figure 5.5. Once again, we see that a higher value of parameter Bi_m implies a higher concentration of the inert species in the pellets, by means of a higher mass transfer through the external surface of the pellets. All results are in agreement with those present in [12], in which the authors discretized the equations using the finite volume method and solved the resulting system of ordinary differential equations using the stiff code DASSL [15].

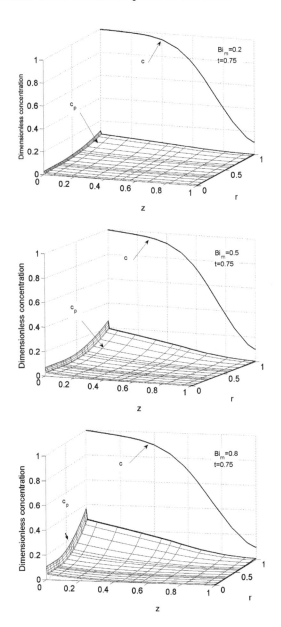

FIGURE 5.5

Fixed-bed adsorption model. Bulk and intraparticle dimensionless concentration profiles in the packed bed, at reduced time instant $t = 0.75$, with $Bi_m = 0.2$ (top), $Bi_m = 0.5$ (middle) and $Bi_m = 0.8$ (bottom).

5.3   Phenol Adsorption on Polymeric Adsorbents

In this section we test the numerical code based on MFEM by considering the mathematical model for adsorption of phenol in a fixed bed of polymeric adsorbent under isothermal operation developed by Costa [13], who presented an experimental system with the purpose of phenol elimination from aqueous solutions. We consider the model for the saturation phase which takes into account the nonlinear Langmuir adsorption equilibrium isotherm, the hydrodynamics described by an axial dispersion flow and the resistances to mass transfer both in the film and inside the particles. The independent variables are the reduced axial coordinate, z, defined by the axial coordinate normalized by the fixed-bed length, L, the normalized radial coordinate for the particle, r, defined by the radial coordinate normalized by the particle radius, R_p and the dimensionless time, t, correspondent to time variable reduced by $\tau(1+\xi)$, with $\tau = L/u_i$, the space time, u_i the interstitial velocity and ξ the adsorbent capacity factor. The model equations for concentration of solute in the fluid phase, C, and for concentration of solute in the particles, C_p, both normalized by the input concentration, C_{in}, can be written in the dimensionless form. The equation that for the mass conservation of phenol in the fluid phase is

$$\frac{\partial c}{\partial t}(z,t) = (1+\xi)\left[\frac{1}{P_e}\frac{\partial^2 c}{\partial z^2}(z,t) - \frac{\partial c}{\partial z}(z,t) - N_f\left[c(z,t) - c_p(1,z,t)\right]\right] \quad (5.15)$$

and the mass conservation of phenol in the adsorbent particle is given by

$$\frac{\partial c_p}{\partial t}(r,z,t) = \frac{(1+\xi)N_D}{\epsilon_p + \frac{K_L\,Q\,\rho_{ap}}{(1+K_L\,C_{in}\,c_p)^2}}\left[\frac{\partial^2 c_p}{\partial r^2}(r,z,t) + \frac{2}{r}\frac{\partial c_p}{\partial r}(r,z,t)\right] \quad (5.16)$$

where K_L and Q are the equilibrium parameters in the Langmuir equation, N_f and N_D are the number of film mass transfer units and the number of intraparticle mass transfer units, respectively, and Pe is the Peclet number. The apparent density is ρ_{ap} and the internal particle porosity is ϵ_p. The N_D and N_f parameters are defined by

$$N_D = \frac{\tau D_{pe}}{R_p^2} \quad (5.17)$$

$$N_f = \frac{1-\epsilon}{\epsilon}a_p k_f \tau \quad (5.18)$$

where D_{pe} is the effective pore diffusivity, $a_p = 3/R_p$ is the particle specific area, k_f is the film mass transfer coefficient, and ϵ is the bed porosity.

The concentrations satisfy the following boundary conditions, $t \geq 0$:

$$c(0, t) = 1 \qquad (5.19)$$

$$\frac{\partial c}{\partial z}(1, t) = 0 \qquad (5.20)$$

$$\frac{\partial c_p}{\partial r}(0, z, t) = 0 \qquad (5.21)$$

and

$$\frac{\partial c_p}{\partial r}(1, z, t) = \frac{1}{3\xi N_D} \frac{K_L Q \rho_{ap}}{1 + K_L C_{in}} N_f \left[c(z, t) - c_p(1, z, t) \right] \qquad (5.22)$$

The initial concentrations for the solute in the fluid phase and for the solute in the pores are, respectively:

$$c(z, 0) = 0 \qquad (5.23)$$

$$c_p(r, z, 0) = 0 \qquad (5.24)$$

Considering the formulation of MFEM for 1D+1d problems, we associate the fixed bed with the macroscale and the particles to the microscale. The functions that links the macroscale and the microscale are, in this case

$$\vartheta(z, t) = (1 + \xi) N_f \, c_p(1, z, t) \qquad (5.25)$$

$$\varpi(z, t) = \frac{1}{3\xi N_D} \frac{K_L Q \rho_{ap}}{1 + K_L C_{in}} N_f \, c(z, t) \qquad (5.26)$$

The first term is calculated and added in the subroutine that determines the residuals of the PDEs while the second term is added to the boundary conditions imposed on the micro equations.

For the numerical simulation, the values used for the model parameters, taken from [13], are listed in Table 5.2

TABLE 5.2
Model parameters. Phenol adsorption on polymeric adsorbents.

K_L	Q	ϵ_p	ρ_{ap}	C_{in}
4.3×10^{-3} l/mg	63.6 mg/g	0.437	537 g/l	100 mg/l

In order to study the influence of the parameters ξ, N_f, N_D, and P_e on the fixed-bed response to a step input of $C_{in} = 100$ mg/l, we consider the model parameters given in Table 5.2 and the four cases presented in Table 5.3. All numerical simulations were done over reduced time interval $[0, 2]$. This

TABLE 5.3

Model parameters. Phenol adsorption on polymeric adsorbents.

	ξ	N_f	N_D	P_e
case (i)	10, 50, 150	50	5	50
case (ii)	50	0.5, 5, 50	5	50
case (iii)	50	50	0.1, 1, 5, 10	50
case (iv)	50	50	5	20, 50, 150

collection of numerical simulations enables us to compare the performance and efficiency of our numerical code based on MFEM to solve 1D+1d problems. The MFEM results are in good accordance with those presented in [13], where double orthogonal collocation in finite elements was implemented and made use of Hermite polynomials.

In the discretization of the macro domain we take a uniform initial grid with 7 finite elements and 10 interior points per element in the macro domain, that implies 78 micro problems coupled to the macro problem. For all the numerical simulations we make use of grids with 5 finite elements for each of the micro domains with 4 interior points in each finite element. The initial mesh nodes for the macro domain are taken equally spaced. To compute the integrals appearing in the equations of the ODE systems we use Lobatto's quadrature with 11 interior quadrature points. For all simulations we use the default MFEM parameters,

$$c_1^{m,j} = c_5^{m,j} = c_6^{m,j} = 10^{-5}, c_2^{m,j} = 0, c_3 = 10^{-3}, c_4 = 10^{-2}$$

Concerning the absolute and relative tolerances of the ODE solver we set $tol_1 = tol_2 = 10^{-5}$. The CPU time per run is less than 2000s.

The effect of the parameters ξ and N_f, over the breakthrough curves is shown in Figure 5.3. The history of solute concentration at the outlet of the fixed bed ($z = 1$) is displayed for case (i), (top) and case (ii) (bottom) of model parameters given in 5.3. Figure 5.3 illustrated the effect of N_D and P_e over the breakthrough curves. The history of solute concentration at the outlet of the fixed bed ($z = 1$) is shown for case(iii) and case (iv). Let us consider for the number of film mass transfer units the values $N_D = 0.1$ and $N_D = 5$, from case (iii). Figure 5.8 gives the dimensionless axial concentration profiles for different instants of time as a function of the reduced axial variable x. For the dimensionless axial concentration profiles at time instants $t = 0, 0.001, 0.04, 0.01, 0.1, 0.2, 0.4, 0.6, 0.8, 1, 1.2, 1.4$ and $t = 2$, Figure 5.8 compares the dimensionless axial concentrations profiles obtained with $N_D = 0.1$ (top) and the dimensionless axial concentrations profiles obtained with $N_D = 5$.

The dimensionless radial concentration profiles in the pores, at the inlet of the fixed bed ($z = 0$) for different instants of time as a function of the reduced axial variable r are also presented. Figure 5.9 displays the results for radial concentration profiles in the pores with $N_D = 0.1$ (top) and $N_D = 5$ (bottom).

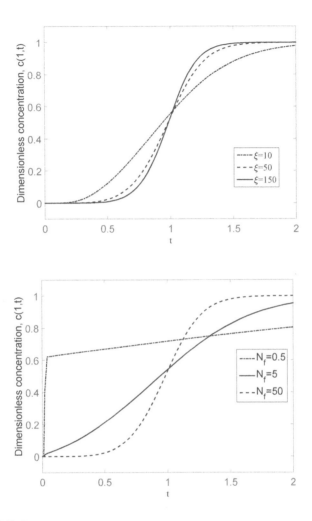

FIGURE 5.6
Phenol adsorption on polymeric adsorbents. Histories of dimensionless solute
concentration at the outlet of the fixed bed as a function of the normalized
time variable: case (i) (top); case (ii) (bottom).

Observing Figure 5.8 and Figure 5.9, we can summarize saying that the axial
profiles are smoother as N_D increase, and the radial profiles at $z = 0$ are
sharper for lower values of N_D. Therefore when N_D increases the resistance
to pore diffusion decreases.

Dimensionless concentration in the particle as a function of the normalized
time, at the exterior surface of the particle ($r = 1$) at the inlet of the fixed

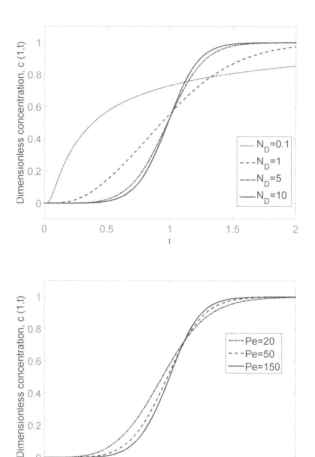

FIGURE 5.7
Phenol adsorption on polymeric adsorbents. Histories of dimensionless solute
concentration at the outlet of the fixed bed as a function of the normalized
time variable: case (iii) (top); case (iv) (bottom).

bed ($z = 0$) and at the outlet of the fixed-bed ($z = 1$), can be found in Figure
5.10, for case (iii) with $N_D = 0.1$ (top) and $N_D = 5$ (bottom).

    Several other numerical simulations were performed with a different num-
ber of finite elements for the mesh associated with the macro domain in order
to investigated the accuracy of the MFEM solution and to accept an MFEM
solution. The methodology for investigating the accuracy of solution consists

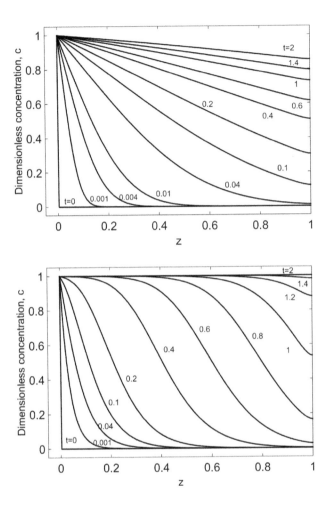

FIGURE 5.8
Phenol adsorption on polymeric adsorbents. Dimensionless axial concentration
profiles at different time instants as a function of the normalized axial variable
z, case(iii) with $N_D = 0.1$ (top) and $N_D = 5$ (bottom).

in an iterative process. First, we fixed the degree of the local approximation;
by example we can use 10 interior points per element. Then we start the
numerical simulation with a certain number of finite elements in the macro
domain. After that a new simulation is performed with more finite elements.
Results from these two runs are compared. If the history of concentrations at
$z = 1$ for the first run is close enough to the history of concentrations at $z = 1$

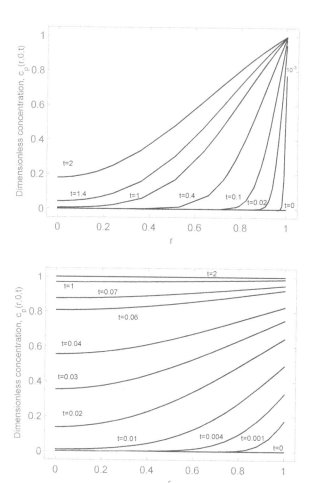

FIGURE 5.9

Phenol adsorption on polymeric adsorbents. Dimensionless radial concentration profiles in the pores at the inlet of the fixed bed ($z = 0$) at different time instants as a function of normalized radial variable r, case(iii) with $N_D = 0.1$ (top) and $N_D = 5$ (bottom).

obtained from the second run, we stop the simulation and present the results given from the first run. If not, the process is repeated. By close enough we mean that the maximum of the absolute difference of history of concentrations at $z = 1$ is smaller than 5×10^{-3}, for example.

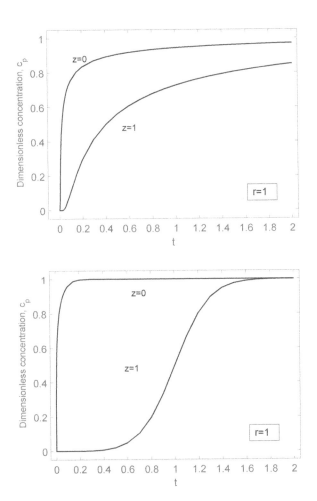

FIGURE 5.10
Histories of dimensionless concentration at the exterior surface of the particle, $r = 1$, at the inlet of the fixed bed ($z = 0$) and at the outlet of the fixed-bed ($z = 1$), case(iii) with $N_D = 0.1$ (top) and $N_D = 5$ (bottom).

5.4   Conclusions

This chapter reports on the moving finite elements method strategy to solve a class of time-dependent partial differential equations presenting a macroscale and a microscale in the 1D+1d structures. The MFEM was successfully used to solve the complex model equations of time-dependent partial differential equations containing two spatial coordinates of different scales, the axial position in the bed and the radial position in the particle.

Bibliography

[1] C. Sereno, A. E. Rodrigues, and J. Villadsen. The moving finite element method with polynomial approximation of any degree. *Computers and Chemical Engineering*, 15(1):25–33, 1991.

[2] C. Sereno, A. E. Rodrigues, and J. Villadsen. Solution of partial differential equations systems by the moving finite element method. *Computers and Chemical Engineering*, 16(6):583–592, 1992. An International Journal of Computer Applications in Chemical Engineering.

[3] Maria do Carmo Coimbra, C. Sereno, and A. E. Rodrigues. Applications of a moving finite element method. *Chemical Engineering Journal*, 84(1):23–29, 2001.

[4] Maria do Carmo Coimbra, C. Sereno, and A. E. Rodrigues. A moving finite element method for the solution of two-dimensional time-dependent models. *Applied Numerical Mathematics*, 44(4):449–469, 2003.

[5] Maria do Carmo Coimbra, C. Sereno, and A. E. Rodrigues. Moving finite element method: Applications to science and engineering problems. *Computer Aided Chemical Engineering*, 14(C):611–616, 2003.

[6] J. D. Rodrigues. *Desenvolvimento do Método de Elementos Finitos Móveis para a simulação de processos*. PhD thesis, Universidade do Porto, 2010.

[7] N. S. Raghavan and D. M. Ruthven. Numerical simulation of a fixed-bed adsorption column by the method of orthogonal collocation. *AIChE Journal*, 29(6):922–925, 1983.

[8] Wangyun Won and Kwang Soon Lee. Adaptive predictive collocation with a cubic spline interpolation function for convection-dominant fixed-bed processes: Application to a fixed-bed adsorption process. *Chemical Engineering Journal*, 166(1):240–248, 2011.

[9] M. Khosravi Koocheksarayi, K. Shams, and Y. Z. Liu. Sorption dynamics in fixed-beds of inert core spherical adsorbents including axial dispersion and langmuir isotherm. *AIChE Journal*, 55(7):1784–1792, 2009.

[10] Zhe Xu, Jian-guo Cai, and Bing-cai Pan. Mathematically modeling fixed-bed adsorption in aqueous systems. *Journal of Zhejiang University SCI-ENCE A*, 14(3):155–176, 2013.

[11] Sahar Afzal, Amir Rahimi, Mohammad Reza Ehsani, and Hossein Tavakoli. Modeling hydrogen fluoride adsorption by sodium fluoride. *Journal of Industrial and Engineering Chemistry*, 16(6):978–985, 2010.

[12] J. P. Barbosa Mota, E. Saatdjian, D. Tondeur, and A. E. Rodrigues. On the numerical solution of partial differential equations with two spatial scales. *Computers Chemical Engineering*, 21(4):387–397, 1997.

[13] C. Costa and A. E. Rodrigues. Design of cyclic fixed-bed adsorption process, part I: Phenol adsorption on polymeric adsorbents. *AIChE Journal*, 31:1645–1654, 1983.

[14] A. C. Hindmarsh. LSODE and LSODIS, two new initial value ordinary differential equation solvers. *ACM-Signum Newslett*, 15:10–11, 1980.

[15] L. R. Petzold. A description of dassl: A differential-algebraic system solver. In R. S. Stepleman et al., editors, *IMACS Transactions on Scientific Computing*, page 65. North-Holland, 1983.

6

Solving Moving Boundary Problems

6.1   Moving Boundary Problems

In this chapter, we present the moving finite element method (MFEM) with piecewise polynomial approximations of arbitrary degree to solve one-dimensional and two-dimensional moving boundary problems or Stefan problems with one or more moving interfaces. For the purpose of increasing the numerical accuracy in the simulation of the important class of Stefan problems, some techniques concerning the treatment of the starting solution when the space domain has an initial zero thickness are developed. The knowledge of the boundary and initial conditions and the pseudo-steady-state solution is applied to find a correct initialization of the numerical model. The techniques investigated are efficient and accurate for solving a large number of problems involving moving boundary interfaces.

Moving boundary problems or Stefan problems can be found in the mathematical modeling of a large range of chemical engineering applications including heat and mass transfer, diffusion, or swelling problems. Due to difficulties in obtaining analytical solutions as reported in Crank [1] numerical methods play an essential role. Stefan problems are difficult to solve because of the presence of one or more moving boundaries at which the phase change occurs. Nevertheless, there has been a growing amount of literature on the development of numerical methods to solve Stefan problems, especially to handle the presence of time-dependent boundary conditions including nonlinear terms. In the literature, it is common to consider two major approaches to Stefan problems. One is the front-tracking method, where the position of the moving boundary is continuously tracked. The other approach uses a fixed-domain formulation that includes enthalpy methods and coordinate transformation. For example, Struckmeier and Unterreiter [2] use a front-tracking process to study the asymptotic behavior of a singular perturbed two-phase Stefan problem with slow diffusion in one of the two phases. Savovic and Caldwell [3] propose the finite difference method to solve one-dimensional Stefan problems with periodic boundary conditions, to determine both the temperature distribution and the position of the moving boundary while Nedoma [4] uses a method based on the weak enthalpy formulation to solve a Stefan problem in Bingham rheology. By considering a similarity solution and a fixed grid enthalpy solution, Lorenzo-Trueba and Voller [5] obtain predictions to the positions of the two moving boundaries in a one-dimensional generalized Stefan problem associated with the formation of sedimentary ocean deltas. Mitchell and Vynnycky [6, 7] derive a numerical algorithm for moving boundary problems using the boundary immobilization method and present a systematic treatment for correct initialization. Studies comparing diverse numerical methods can be found in the literature, [8], [9], and [10].

6.1.1   MFEM and moving boundary problems in 1D: Strategies for choosing a starting solution

In order to solve the moving boundary problems with nonlinear time-dependent multiphase boundary conditions, we develop an algorithm based on the moving finite element method (MFEM) with higher-order Lagrange basis functions for both the trial and the test spaces [11]. Revisiting the study presented by Robalo [12, 13], we extend to a more general process when two or more internal moving boundaries are found. Furthermore, we focus our attention on methodologies for determining the starting solution when applying the MFEM to solve this kind of problem with initial zero length spatial domains.

Let us consider one-dimensional moving boundary problems. The proposed formulation of the MFEM allows the use of different spatial meshes, on a fixed domain, each one being associated with a different dependent variable. Making use of this characteristic of the MFEM formulation, it is possible to decompose the spatial domain, Ω, by using spatial nodes describing the positions of the internal moving boundary to be tracked. Let n be the number of moving interfaces that split the domain Ω into $n+1$ intervals, D_i, $i = 1, \ldots, n+1$,

$$\Omega = \cup_{i=1}^{n+1} D_i \tag{6.1}$$

The position of the interface S_i is time dependent, connecting two neighborhood domains D_i and D_{i+1} at each instant. Thus, D_i is also time dependent. To each of these subdomains is associated a set of partial differential equations, initial and boundary conditions and equations describing the velocity of the moving boundary. Usually we refer to phase i of the problem when describing the problem over the spatial sub domain D_i. If $\mathbf{y}$ is the vector of all dependent variables and $t = 0$ is the initial time, the moving boundary problem can be described by a system of partial differential equations with the general form

$$\frac{\partial y_m^i}{\partial t} = F_{1,m}^i \left(x, t, \mathbf{y}, \frac{\partial \mathbf{y}}{\partial x} \right) \frac{\partial^2 y_m}{\partial x^2} + F_{2,m}^i \left(x, t, \mathbf{y}, \frac{\partial \mathbf{y}}{\partial x} \right) \tag{6.2}$$

where $x \in D_i$, $t \geq 0$ and subject to the initial conditions,

$$y_m^i(x, 0) = g_m^i(x). \tag{6.3}$$

Robin or Dirichlet time-dependent boundary conditions in the fixed endpoints of Ω are allowed, so we can have one of the following boundary conditions,

$$\frac{\partial y_m^i}{\partial x} = \alpha_m^i(t)y_m^i + \beta_m^i(t) \tag{6.4}$$

$$y_m^i = \gamma_m^i(t) \tag{6.5}$$

with real time-dependent functions α_m^i, β_m^i e γ_m^i.

At each interface position, S_i, a more general nonlinear condition, can be used. This general condition is defined by an equation of the type,

$$h_m^i \left(\mathbf{S}, t, \mathbf{y}^i\big|_{S_i}, \mathbf{y}^{i+1}\big|_{S_i}, \frac{\partial \mathbf{y}^i}{\partial x}\bigg|_{S_i}, \frac{\partial \mathbf{y}^{i+1}}{\partial x}\bigg|_{S_i} \right) = 0 \tag{6.6}$$

Here $\mathbf{S}$ is the vector of all interface positions and $\mathbf{y}^i$ and $\mathbf{y}^{i+1}$ are subvectors of $\mathbf{y}$ associated to D_i and D_{i+1}, respectively. This function h_m^i depends on all interface positions. Finally we assume that the interface movement is described by an equation of the form,

$$\frac{dS_i}{dt} = v \left(\mathbf{S}, t, \mathbf{y}^i\big|_{S_i}, \mathbf{y}^{i+1}\big|_{S_i}, \frac{\partial \mathbf{y}^i}{\partial x}\bigg|_{S_i}, \frac{\partial \mathbf{y}^{i+1}}{\partial x}\bigg|_{S_i} \right) \tag{6.7}$$

When solving moving boundary problems using the MFEM, it is important to be able to deal with two different situations that may occur. In fact, there is a possibility that one of the domains, D_i has null measure at some instant of time, which is the case of two interfaces with opposite movement. In this situation, we have to stop the simulation when the distance between the two interfaces is close to zero, and the user must supply this value. If initially the interface is at one of the fixed boundaries, then a more elaborate procedure is necessary. In fact, to start the MFEM simulation, two challenging issues must be addressed: firstly the need for a non-zero thickness spatial domain and secondly the necessity of considering a similar governing equation to the original problem whose solution is known a priori.

To overcome the case of having a zero length spatial domain at $t = 0$, a small starting time t_0 must be obtained, corresponding to a similar problem to the original one. Therefore it is necessary to set the position of the interface at time t_0, $s(t_0)$, arbitrarily close to the real one, $s(0)$, and also to define an approximate solution at time instant t_0. To accomplish that, and depending on the problem under study, the mathematical model, boundary conditions, initial conditions, and eventually a pseudo-steady-state solution, should be considered to produce an accurate starting solution.

One possible approach is to use the information you can get from the physical problem and deduce the time instant t_0 and the starting solution from (linear) approximations of the boundary and initial conditions. For other problems where the pseudo-steady-state solution is known, we may choose an arbitrarily small initial instant t_0 and use the knowledge of the pseudo-steady-state solution to adjust the initial location of the interface and the starting solution. Moreover, it might be necessary to extend the original problem to an interval large enough to contain the positions of the moving interface during the process since the MFEM demands a fixed spatial domain. For each of the problems presented in this chapter, in Section 6.2 to Section 6.6 we discuss the procedure used for choosing the starting solutions. The numerical simulations with the MFEM demand small penalty constants defined by Equation 2.40. For the 1D moving boundary problems presented in this chapter the default penalty constants are

$$c_1^{m,j} = 10^{-5}, \ c_2^{m,j} = 0, \ c_3^{m,j} = 10^{-3}, \ c_4^{m,j} = 10^{-2}, \ c_5^{m,j} = c_6^{m,j} = 10^{-5}$$

and the default ODE tolerances are $tol_1 = tol_2 = 10^{-9}$.

The numerical code used to implement the MFEM for 1D moving boundary problems is written in FORTRAN and the integration of the ODE system of the MFEM equation is accomplished with the LSODIS routine [14].

6.1.2   MFEM and moving boundary problems in 2D

The MFEM for time-dependent problems in two-dimensional spatial domains is presented in Section 2.3.1 of Chapter 2. Solving moving boundary problems in two-dimensional spatial domains is a hard task for numerical methods and we took up the challenge of developing the 2D-MFEM for solving problems of this type. Let Ω be the spatial domain. The coordinates of a point in Ω are denoted by (x_1, x_2).

First of all, and in a different way to what has been done before, we assume that the spatial domain of the problem may have a non-rectangular geometry, an irregular geometry, and that a different spatial mesh can be used to represent each dependent variable. As previously, the spatial domain is discretized using triangular finite elements. The discretization of irregular fixed boundaries is an added source of error and a large number of nodes may be required to minimize errors. Figure 6.1 shows the discretization of a region between two circles. As Figure 6.1 illustrates the shaded region is not part of the computational domain. Errors of this kind are inherent to the discretization and can be minimized by using a larger number of points or, for certain problems, by taking advantage of the symmetry of the problem as has been seen in Chapter 4. In fact, different types of symmetry may be used. If it is known that a 2D model has symmetric solutions with respect to the x_1 component, then it is desirable that the mesh used to discretize the space domain has the same property. Figure 6.2 compares two meshes with the same number of nodes. The mesh represented on the right is more favorable for obtaining accurate solutions because it reflects the $\mathcal{O}x_1$ symmetry of the solution.

For moving boundary problems in the 2D spatial domain, following the ideas developed for 1D problems, the spatial domain is decomposed as Equation 6.1 shows, but now the interface is a line. The boundary of the spatial domain at the time instant t is the union of the fixed boundary, Γ_F, and the

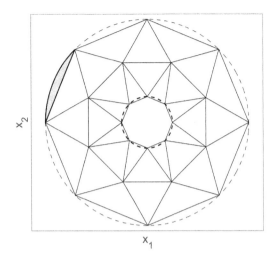

FIGURE 6.1
Moving boundary problems. Space domain decomposition. Discretization of
an irregular 2D spatial domain.

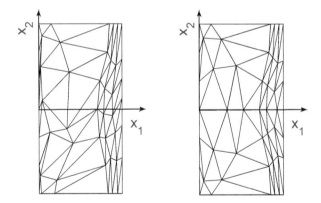

FIGURE 6.2
Moving boundary problems. Space decomposition of a regular 2D spatial domain. Non-preserving symmetry (left) and preserving $\mathcal{O}x_1$ symmetry (right).

moving boundary, Γ_M, which may include one or more interfaces. In each
phase the spatial domain must be discretized and different meshes associated with each dependent variable should be obtained. The spatial nodes that

belong to the fixed boundary are allowed to move according to the characteristics of the fixed boundary. To discretize the moving boundary, a set of spatial nodes $\mathbf{X}_{S_j}$ in the interface is used. These nodes are included in the spatial mesh. Figure 6.3 shows a piecewise linear approximation to the moving boundary and the normal outward unit vector to the interface in phase 1. The movement of the nodes that belong to the moving boundary is determined by the velocity of this boundary along the normal vector.

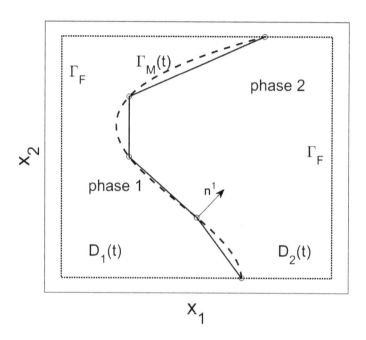

FIGURE 6.3
Moving boundary problems. Space domain decomposition. Interface discretization and outward normal unit vector to the moving interface in phase 1.

The coordinates of the nodes belonging to the moving boundary are denoted by $\mathbf{X}_{s_j} = \left((X_1)_{s_j}, (X_2)_{s_j} \right)^T \in \Gamma_M(t)$. Assuming a null tangential velocity at the nodes that belong to the moving boundary, the speed of the moving boundary is given by

$$\sum_{I=1}^{2} n_I|_{\mathbf{X}_{s_j}} (\dot{X}_I)_{s_j} = -\sum_{I=1}^{2} n_I|_{\mathbf{X}_{s_j}} \left.\frac{\partial U}{\partial x_I}\right|_{\mathbf{X}_{s_j}} \tag{6.8}$$

where $n_I|_{\mathbf{X}_{s_j}}$ is the I-th coordinate of the normal vector to the moving boundary, calculated at $\mathbf{X}_{s_j}$. Two relevant questions arise at this point: First, how to define the approach of the normal vector to Γ_M at each time instant? Secondly, how to compute the spatial derivatives present in Equation 6.8 in a mesh vertex that belongs to Γ_M. For the first one, we consider the normalizing vector sum of the external normal vectors to the two nodal common edges of the interface. For the second one, we define the spatial derivatives as the average of the local finite element derivatives. Equation 6.8 associated with $(X_I)_{s_j}$ is external to the MFEM equations and must be added to the resulting ODE system. Depending on the problem to be solved, it may be necessary to include other equations if the movement of $\mathbf{X}_{s_j}$ is established a priori. In this chapter two numerical examples are presented and solved with the MFEM. The first example, the Frank spheres problem in Section 6.7, is an excellent test that we perform to reproduce the exact solution with the code based on the MFEM. This moving boundary problem is posed in a 2D circular space domain with only one active phase and the moving boundary has a radial motion. The second problem in Section 6.8 is a solidification problem with transition temperature range and so is a problem having two moving interfaces. The solid region is separated from the mushy region by a solid-mush moving boundary and the mushy region is separated from the liquid region by a mushy-liquid region. For the MFEM implementation we must define the problem in three phases. Initially, in phase 1 the problem has a spatial domain with null measure and, at the end of the computations in phase 3, the spatial domain also had null measure. Numerical computations show the feasibility and the efficiency of the moving finite element method in handling moving boundary problems in the 2D space domain.

Simplified penalty constants defined by Equation 2.40 are used. The user must supply only three constants, $c_1^{m,j}$, $c_3^{m,j}$, and $c_5^{m,j}$. For 2D problems $c_5^{m,j}$ represents the minimum admissible area for a finite element and, as before, $c_1^{m,j}$ relates to spring forces and $c_3^{m,j}$ concerns viscosity function. The other constants are set to

$$c_2^{m,j} = 0, \ c_4^{m,j} = c_3^{m,j}, \ c_6^{m,j} = c_5^{m,j}$$

The numerical code used to implement MFEM for 2D moving boundary problems is written in MATLAB$^{\circledR}$ [15] and the integration of the ODE system of MFEM equation is accomplished by `ode15s` MATLAB routine.

6.2 Melting and Freezing Problems

A melting problem. One moving interface model.

Consider the melting of a certain material that is initially at its melting temperature, T_M, on the spatial domain $[0, +\infty[$ and then is heated at $z = 0$

for time $\theta > 0$ causing it to melt. A boundary, at which melting occurs, moves from $z = 0$ to the right, and separates a region where the material is liquid from one where the material is solid at melting temperature. If T denotes the temperature in the liquid region at time θ, and $s(\theta)$ denotes the position of the interface at time θ, the melting process may be described by

$$c_p \rho \frac{\partial T}{\partial \theta} = K \frac{\partial^2 T}{\partial z^2}, \quad 0 < z < s(\theta), \ \theta > 0 \tag{6.9}$$

where the constants c_p, ρ, K are the specific heat, density, and conductivity, respectively, subject to the fixed boundary condition

$$T(0,t) = T_0, \ t \geq 0 \tag{6.10}$$

where $T_0 > T_M$ is a constant temperature. Initial conditions are

$$T(z,0) = T_M, \ z > 0$$
$$s(0) = 0 \tag{6.11}$$

Two further conditions are needed on the moving interface

$$T(s(\theta), t) = T_M, \ t \geq 0 \tag{6.12}$$

$$\rho L s'(\theta) = - K \left. \frac{\partial T}{\partial z} \right|_{z=s(\theta)} \tag{6.13}$$

where the constant L is latent heat of melting. Equation 6.13 is known as the *Stefan condition* and it expresses the heat balance on the interface. All variables and the heat parameters denote quantities expressed in physical units. To make the solution more meaningful and simpler, it is more convenient to work with non-dimensional variables. Thus, introducing the new variables

$$x = \frac{z}{z_{ref}} \tag{6.14}$$

$$S = \frac{s}{z_{ref}} \tag{6.15}$$

$$t = \frac{K}{c_p \rho} \frac{\theta}{z_{ref}^2} \tag{6.16}$$

$$u(x,t) = \frac{T - T_M}{T_0 - T_M} \tag{6.17}$$

where z_{ref} is some standard length, reduces the model to the following equations

$$\frac{\partial u}{\partial t} = \frac{\partial^2 u}{\partial x^2} \tag{6.18}$$

with $0 < x < S(t)$, $t > 0$ and subject to the initial and boundary conditions

$$u(x, 0) = 0 \tag{6.19}$$
$$S(0) = 0 \tag{6.20}$$
$$u(S(t), t) = 0 \tag{6.21}$$
$$u(0, t) = 1 \tag{6.22}$$

and

$$\frac{dS}{dt} = -\frac{1}{\beta} \frac{\partial u}{\partial x}\bigg|_{x=S(t)} \tag{6.23}$$

where $\beta = L/(c_p(T_0 - T_M))$ is the dimensionless latent heat parameter, that corresponds to the reciprocal of the Stefan number.

To obtain the numerical solution with the MFEM algorithm, instead of the half plane, we should consider a limited space domain, $x \in [0, \ell]$, with ℓ large enough to contain the positions of the moving interface during the melting process. The spatial domain is decomposed into two subdomains by using the interface position, $[0, \ell] = [0, S(t)] \cup [S(t), \ell]$. The temperature needs to be found only for $x \in [0, S(t)]$. This Stefan problem has two phases but only phase 1 is an active phase. Phase 2 can be seen as an inactive phase and the problem is called a one-phase Stefan problem since only one active phase is considered as active in the melting process. Nevertheless, to apply the MFEM we need to define this inactive phase. Another issue that arises in applying the MFEM is that in phase 1, at $t = 0$, the space region has zero thickness, and hence a starting solution must be calculated, in phase 1.

Concerning the application of the MFEM, let us assume that $y_1 = u(x, t)$ is the dimensionless temperature variable in phase 1 and $S = S(t)$ is the position of the interface at time t. The dimensionless model for phase 1 is described by Equation 6.18, with the initial conditions defined by Equation 6.19 and Equation 6.20. The boundary condition at the interface is given by Equation 6.21 and Equation 6.23 is the Stefan condition giving a description of the velocity of the interface. With regard to the fixed boundary condition at $x = 0$, Equation 6.22, four different problems are analyzed, corresponding to the following four conditions, similar to the work by Mitchell and Vynnycky [6].

$$y_1(0, t) = 1 \quad (i)$$
$$y_1(0, t) = \exp(t) - 1 \quad (ii)$$
$$\frac{\partial y_1}{\partial x}(0, t) = -\exp(t) \quad (iii) \tag{6.24}$$
$$y_1(0, t) = 1 + \epsilon \sin(\omega t) \quad (iv)$$

The dimensionless temperature in phase 2 is denoted by y_2 and the governing equation for a null solution is:

$$\frac{\partial y_2}{\partial t} = \frac{\partial^2 y_2}{\partial x^2}, \quad S(t) < x < \ell, t > 0 \tag{6.25}$$

with null initial and boundary conditions,

$$y_2(x, 0) = y_2(S(t), t) = y_2(\ell, t) = 0 \tag{6.26}$$

In addition, the user must define some parameters. For the four melting problems in the study, problems (*i-iv*), the same set of MFEM parameters is used. We consider 3 finite elements in each phase, with 10 interior points in phase 1 and 5 interior points in phase 2, in each finite element. Lobatto's quadrature with 33 interior quadrature points was chosen to compute all the integrals appearing in the equations of the ODE systems. Default values for the MFEM penalty functions and ODE solver tolerances were considered. For phase 1 the initial node positions are

$$X_{1,1} = 0, X_{1,2} = 3 \times 10^{-5}, X_{1,3} = 5 \times 10^{-5}, X_{1,4} = S(t_0) \tag{6.27}$$

and, for phase 2,

$$X_{2,1} = S(t_0), X_{2,2} = 10^{-3}, X_{2,3} = 5 \times 10^{-3}, X_{2,4} = \ell \tag{6.28}$$

Numerical computations take less than 6 seconds.

As mentioned before, the space region has zero thickness at $t = 0$, and hence a starting solution must be calculated in phase 1, for each one of the problems (*i-iv*). It is required to determine the initial time t_0 and the initial condition $y_1(0, t_0)$, assuming that the phase change starts at instant t_0 and that the position of the interface $S(t_0)$ is arbitrarily close to the left end of the physical space domain. In order to find the initial instant t_0 for the problems defined by Equations (6.18)–(6.24-*i-iv*), the following approximation from Equation 6.23 is considered

$$\beta \frac{S(t_0) - 0}{t_0 - 0} = -\frac{y_1(S(t_0), t_0) - y_1(0, t_0)}{S(t_0) - 0}. \tag{6.29}$$

According to the boundary condition at the interface, Equation 6.21, t_0 is chosen to satisfy the equation,

$$\beta[S(t_0)]^2 = t_0 y_1(0, t_0), \tag{6.30}$$

with $S(t_0)$ small enough. For all the numerical simulations $S(t_0)$ was set to $S(t_0) = 10^{-4}$.

For problem (*i*) the boundary condition at $x = 0$ is $y_1(0, t) = 1$, so the value of t_0 is determined by Equation 6.30, $\beta[S(t_0)]^2 = t_0$. Therefore for the numerical simulation $t_0 = \beta \times 10^{-8}$ and, in the first phase, it is assumed that the starting solution ($t = t_0$) is linear decreasing from $y_1 = 1$ at $x = 0$ to $y_1 = 0$ at $x = S(t_0)$. For the MFEM implementation, we consider $\ell = 3$ to well accommodate the movement of the interface inside the spatial domain of phase 2.

The exact solution of the Stefan problem (i) in a semi-infinite slab is known [1, 10],

$$y_1(x,t) = 1 - \frac{erf[x/(2\sqrt{t})]}{erf(\lambda)} \tag{6.31}$$

$$S(t) = 2\lambda\sqrt{t} \tag{6.32}$$

where erf denotes the error function, $erf(x) = \frac{2}{\pi}\int_0^x exp(-s^2)ds$, and λ satisfies the implicit equation,

$$\sqrt{\pi}\,\beta\,\lambda\,erf(\lambda)\,\exp(\lambda^2) = 1. \tag{6.33}$$

Hence the error of the MFEM numerical solution can be computed.

Figure 6.4 presents the results of the numerical simulations of problem (i) for two different values of β parameter, $\beta = 0.2$ and $\beta = 2.0$. The evolution of the reduced position of the interface in time is shown on the left and, on the right, the absolute error in the computation of the position of the interface $s(t)$ is presented in a logarithm scale. The absolute error in the position of the moving interface is less than 9.5×10^{-6} when $\beta = 0.2$ and less than 3.2×10^{-5} when $\beta = 2.0$.

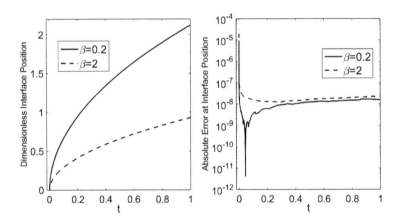

FIGURE 6.4
A melting problem. Dimensionless interface position versus reduced time (left) and evolution of absolute error at the interface position with time (right) for problem (i) with $\beta = 0.2$, solid, and $\beta = 2$, dashed.

For problem (ii), in order to find an accurate t_0 that satisfies $S(t_0) = 10^{-4}$, Equation 6.30 leads to the following relation,

$$10^{-8}\,\beta = t_0\,(exp(t_0) - 1). \tag{6.34}$$

Taking $\beta = 1$, then $t_0 = 10^{-4}$ and $y_1(0, t_0) = 10^{-4}$. Therefore, in the first phase the starting solution is linear, decreasing from $y_1 = 10^{-4}$ at $x = 0$ to $y_1 = 0$ at $x = S(t_0)$. Considering $\beta = 1$, both problems corresponding to conditions (ii) or (iii) have the same analytical solution [10]

$$y_1(x, t) = \exp(t - x) - 1 \quad \wedge \quad S(t) = t. \tag{6.35}$$

However, for problem (iii) the value of $y_1(0, t_0)$ is not known. From the value of the derivative $\frac{\partial y_1}{\partial x}(0, t) = -\exp(t)$ it is possible to establish the approximation

$$-exp(t_0) = \frac{y_1(S(t_0), t_0) - y_1(0, t_0)}{S(t_0) - 0}. \tag{6.36}$$

Thus $y_1(0, t_0) = S(t_0) \exp(t_0)$ and, by Equation 6.29 the relationship defining t_0 can be defined as

$$\beta S(t_0) = t_0 \exp(t_0), \tag{6.37}$$

with $S(t_0)$, arbitrarily small. If $\beta = 1$, then $t_0 = 10^{-4}$ and $y_1(0, t_0) = 10^{-4}$, and in the first phase the starting solution is linear, decreasing from $y_1 = 10^{-4}$ at $x = 0$ to $y_1 = 0$ at $x = S(t_0)$.

For problem (ii) the absolute error in the position of the interface over time was less than 4.9×10^{-9} and the absolute error of the temperature at $t = 1$ was less than 2.2×10^{-9}. Problem (iii) has the same analytical solution as problem (ii), but smaller errors were obtained. In problem (iii), the absolute error in the position of the interface over time was less than 9.6×10^{-11} and the absolute error in dimensionless temperature at $t = 1$ was less than 1.3×10^{-10}. Figure 6.5 shows, for both problems with $\beta = 1$, the evolution of the moving boundary through time (on the left) and, the absolute error of the interface position (on the right). Figure 6.6 presents the dimensionless temperature profile on the left and, on the right the absolute error for dimensionless temperature at the instant $t = 1$ as a function of dimensionless space, for both problems. For numerical simulations we take $\ell = 2$.

Finally, concerning problem (iv), the analytical solution is not known but a starting solution can be obtained by the same procedure. Let $\epsilon = -0.5$ and $\omega = \frac{\pi}{2}$. Instant t_0 is again obtained from Equation 6.29:

$$10^{-8} \beta = t_0 \left[1 - 0.5 \sin \left(\frac{\pi}{2} t_0 \right) \right]. \tag{6.38}$$

In phase 1 the starting solution is linear decreasing from $y_1 = 1 - 0.5 \sin \left(\frac{\pi}{2} t_0 \right)$ at $x = 0$ to $y_1 = 0$ at $x = S(t_0)$.

The numerical results with $\ell = 10$ are illustrated in Figure 6.7 and Figure 6.8. Figure 6.7 shows the evolution of the moving boundary for $\beta = 0.5$, $\beta = 1$, and $\beta = 5$ with initial instants $t_0 t_0 = 5 \times 10^{-9}$, $t_0 = 10^{-8}$, and $t_0 = 5 \times 10^{-8}$, respectively, obtained from Equation 6.38. Figure 6.8 shows some comparative dimensionless temperature profiles for $t = 4$ to $t = 8$ with $\beta = 1$ as a function of the reduced space, normalized by $S(t)$, that is, normalized by the position of the interface at the time instant at which the profile is taken.

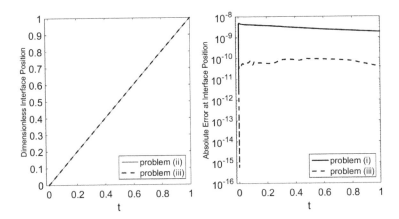

FIGURE 6.5
A melting problem. Dimensionless interface position versus reduced time (left) and evolution of absolute error at the interface position with time (right) for problem (ii), solid, and for problem (iii), dashed, with $\beta = 1$.

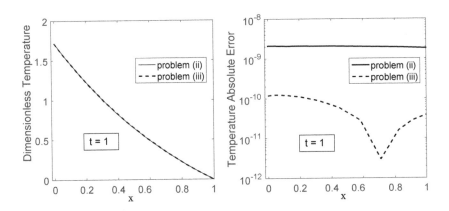

FIGURE 6.6
A melting problem. Dimensionless temperature at reduced time $t = 1$ as a function of reduced space x (left) and absolute error of dimensionless temperature at $t = 1$ as a function of x (right) for problem (ii), solid, and for problem (iii), dashed, with $\beta = 1$.

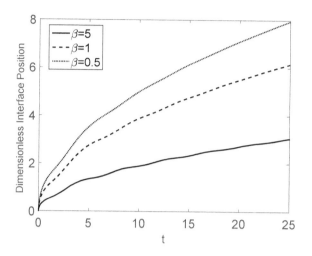

FIGURE 6.7
A melting problem. Dimensionless interface position as a function of dimensionless time t for different values of β parameter, $\beta = 0.5$, $\beta = 1$, $\beta = 5$, problem (iv).

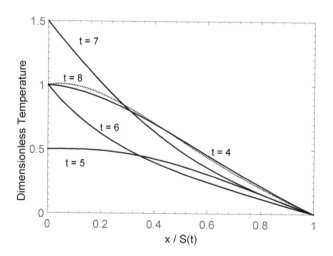

FIGURE 6.8
A melting problem. Dimensionless temperature profiles as a function of $x/S(t)$ for several different instants of dimensionless time, problem (iv) with $\beta = 1$.

Notice that from the boundary condition imposed at $x = 0$ we obtain,

$$\frac{\partial y_1}{\partial t} = -\frac{\pi}{4} \cos\left(\frac{\pi}{2} t\right); \quad t \geq t_0. \tag{6.39}$$

Thus, looking at the governing heat equation, Equation 6.18, it may be observable that the sign of the derivative that appears in Equation 6.39 influences the concavity of the temperature graph near $x = 0$. Figure 6.8 is in conformity with this fact. For example, at $t = 7$ the derivative is equal to zero and the concavity of the graph is almost zero. Taking another time instant, for example, $t = 8$, the derivative is negative, which is in accordance with the concavity of the temperature graph. If, for example, $t = 6$, then the derivative is positive, in conformity with the concavity of the temperature graph.

The results achieved demonstrate that the MFEM and the adopted technique using boundary conditions and initial conditions to start the simulations in the case of a starting domain of zero thickness, lead to accurate numerical solutions, compared with those presented in [6] where finite-difference discretization schemes were used along with the boundary immobilization method.

A freezing problem. Two moving interfaces model.

Having solved the classical Stefan problem for modeling the melting of ice, we now present the problem of a liquid solidification with the presence of two moving interfaces resulting from heat transfer. The spatial domain is a limited interval with three regions, ice-water-ice. The solid phases (ice) grow with respect to the liquid phase (water). The liquid is initially at its freezing temperature T_f, and is in contact with a surface maintained at a lower temperature. Both ends of the spatial domain are at a constant temperature T_0, with $T_0 < T_f$. Let us consider the dimensionless model as in the previous example. Figure 6.9 presents a scheme of the dimensionless spatial domain at reduced time t. During the solidification process, interface S_1 moves to the right while interface S_2 moves to the left, toward each other. In order to com-

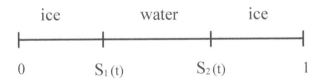

FIGURE 6.9
A freezing problem with two interfaces. Ice-water-ice problem with two moving interfaces $S_1(t)$ and $S_2(t)$.

pare with results from Djomehri and George [16] we consider the dimensionless temperature, $u(x, t) = (T - T_f)/(T_f - T_0)$.

Let y_i be the dimensionless temperature in the i-phase, x the reduced space variable, and $S_1(t)$ and $S_2(t)$ the reduced positions of the interfaces at the reduced instant time t. The dimensionless partial differential equations that describe the problem are

$$\frac{\partial y_1}{\partial t} = \frac{\partial^2 y_1}{\partial x^2}, \ 0 < x < S_1(t), \ t > 0$$

$$\frac{\partial y_2}{\partial t} = 0, \ S_1(t) < x < S_2(t), \ t > 0 \qquad (6.40)$$

$$\frac{\partial y_3}{\partial t} = \frac{\partial^2 y_3}{\partial x^2}, \ S_2(t) < x < 1, \ t > 0$$

and the dependent variable temperature should satisfy the following boundary conditions:

$$y_1(0, t) = -1$$
$$y_1(S_1(t), t) = y_2(S_1(t), t) = 0$$
$$y_2(S_2(t), t) = y_3(S_2(t), t) = 0 \qquad (6.41)$$
$$y_3(1, t) = -1$$

The interfaces move according to the velocities

$$\frac{dS_1}{dt} = \frac{\partial y_1}{\partial x}\Big|_{(S_1(t), t)}$$

$$\frac{dS_2}{dt} = \frac{\partial y_3}{\partial x}\Big|_{(S_2(t), t)} \qquad (6.42)$$

Finally, we consider the nonuniform initial conditions, [16]

$$y_1(x, 0) = 4x - 1$$
$$y_2(x, 0) = 0$$
$$y_3(x, 0) = -4x + 3 \qquad (6.43)$$
$$S_1(0) = 0.25$$
$$S_2(0) = 0.75$$

In all the numerical simulations we use 4 finite elements decomposing each phase and 2 interior points per finite element. The initial positions of the separation nodes are

$$X_{1,1} = 0, X_{1,2} = 0.1, X_{1,3} = 0.15, X_{1,4} = 0.2, X_{1,5} = S_1(t_0) \qquad (6.44)$$

for phase 1, and

$$X_{2,1} = S_1(t_0), X_{2,2} = 0.3, X_{2,3} = 0.5, X_{2,4} = 0.7, X_{2,5} = S_2(t_0) \qquad (6.45)$$

for phase 2, and

$$X_{3,1} = S_2(t_0), X_{3,2} = 0.8, X_{3,3} = 0.85, X_{3,4} = 0.9, X_{3,5} = 1 \qquad (6.46)$$

for phase 3.

With regard to the MFEM parameters, default values are used. However two different simulations are considered to illustrate the effect of the penalty constant $c_1^{m,j}$ on the movement of the spatial nodes. It takes approximately 3 seconds to compute each simulation. Due to the impossibility of zero-thickness in the second phase, we impose a common limit on the movement of the two interfaces. In this case and for the sake of symmetry, we choose a limit at $x = 0.5$. The integration process should stop when one of the two interfaces reaches this limit value. Computations stop at reduced time $t = 0.121$. At this time instant, $t = 0.121$, the interface positions are $S_1 = 0.4997$ and $S_2 = 5.003$, respectively.

Next, we present the numerical results achieved, which are in good agreement with those presented in [16]. Figure 6.10 shows the dimensionless temperature profiles along dimensionless time and over the normalized space domain.

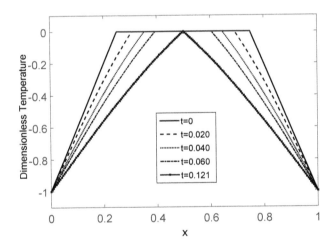

FIGURE 6.10
A freezing problem with two interfaces. Dimensionless temperature profiles as function of reduced space, x, for the reduced time instants $t = 0$, $t = 0.2$, $t = 0.4$, $t = 0.6$ and $t = 0.121$.

In Figure 6.11, we can observe the grid evolution over the normalized space domain. There are two fixed nodes at the fixed boundaries, and the movement of the two interfaces is given by the solid lines converging to $x = 0.5$. The meshes nodes move to follow the moving interfaces, ensuring a good representation of the solution. The numerical simulation performed with different

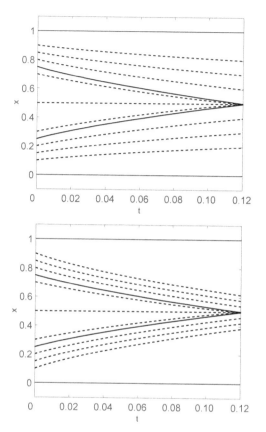

FIGURE 6.11

A freezing problem with two interfaces. Meshes evolution with reduced time. Solid lines refer to the fixed boundaries, $x = 0$, $x = 1$ and to the moving interfaces. Simulation performed with default penalty constants, (top) and with different penalty constants, $c_1^{1,1} = c_1^{3,4} = 10^{-2}$, (bottom).

penalty constants, $c_1^{1,1} = c_1^{3,4} = 10^{-2}$, produces distinct meshes, as Figure 6.11 (bottom) shows. An adjustment in the values of the penalty constant c_1 in the first finite element of the first phase and in the last finite element of the third phase results in alterations in the node movements but does not interfere with the accuracy of the solution. Consequently, the temperatures obtained with the parameters related to Figure 6.11 (bottom) are the same as those obtained with the parameters used in the simulation that leads to Figure 6.11 (top). The increase of this small constant c_1 implies that the nodes move closely to the interfaces. Remember that, in the MFEM formulation, the only influence of the penalty functions is on the node movement, preventing singularities in the mass matrix and avoiding the collapse of the spatial grids.

6.3 The Shrinking Core Model

The shrinking core model is used to describe phenomena in which solid particles are being transformed and, as a result, the amount of the material reacting is *shrinking*. We are interested in solving the problem of fluid-solid non-catalytic heterogeneous reactions by the shrinking core model. This problem was described by Carabin et al. [17] and by Robalo [12]. The effects of transient diffusion, nonlinear kinetics and nonisothermal scenarios are studied. Let us consider the reaction of fluid species A with spherical particles of solid S with radius R_p at time $t = 0$. There is diffusion of reactant A through the porous layer of product P, reaction at interface of solid S, producing porous solid P and product B that diffuses out through the solid. As the reaction progresses, the interface between the solid and the porous layer with radius r_c moves inward the particle, as Figure 6.12 illustrates. If ϑ_i are the stoichiometric coefficients, the reaction of the shrinking core takes the form:

$$\vartheta_A\, A_{(f)} + \vartheta_S\, S_{(s)} \quad \rightarrow \quad \vartheta_B\, B_{(f)} + \vartheta_P\, P_{(s)} \tag{6.47}$$

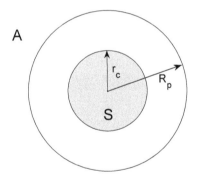

FIGURE 6.12
The shrinking core model. Diffusion through a spherical pellet.

The shrinking core model. Isothermal case.

We start with the isothermal case and we care about the effect of diffusion and the kinetics of the reaction at the interface. Let $c(x, t)$ be the dimensionless concentration of reactant A, normalized by the bulk molar concentration of A, C_{Ab}, where x is the radial position, r, normalized by R_p, and t is the reduced time by the characteristic time, $t_c = R_p^2/D$. The effective diffusivity of A is D, and $x_c(t)$ is the reduced position of the interface of the solid unreacted core. At reduced time instant $t = 0$ we have $x_c(0) = 1$.

The equation for the diffusion of reactant A through the particle, in dimensionless form, is

$$\frac{\partial c}{\partial t} = \frac{\partial^2 c}{\partial x^2} + \frac{2}{x}\frac{\partial c}{\partial x}; \quad x_c(t) < x < 1, t > 0 \tag{6.48}$$

In order to adapt this model to our MFEM formulation for Stefan problems, we extend the spatial domain to the fixed interval $[0,1]$, so it becomes a biphasic problem with the two phases separated by the interface $x_c(t)$. In the first phase we introduce equations similar to those in the second phase, so that the solution for the first phase will be the null solution. Instead of the notation c for the concentration, let y_1 and y_2 be the dependent variables that represent the concentration of reactant A in the first and second phases, respectively. So the diffusion equations over all the domain could be:

$$\frac{\partial y_1}{\partial t} = \frac{\partial^2 y_1}{\partial x^2} + \frac{2}{x}\frac{\partial y_1}{\partial x}; \quad 0 < x < x_c(t), t > 0 \tag{6.49}$$

$$\frac{\partial y_2}{\partial t} = \frac{\partial^2 y_2}{\partial x^2} + \frac{2}{x}\frac{\partial y_2}{\partial x}; \quad x_c(t) < x < 1, t > 0 \tag{6.50}$$

subject to the boundary conditions,

$$y_1(0,t) = 0 \tag{6.51}$$

$$y_1(x_c(t),t) = 0 \tag{6.52}$$

$$\left.\frac{\partial y_2}{\partial x}\right|_{(x_c(t),t)} = Da\, y_2^n(x_c(t),t) \tag{6.53}$$

$$\left.\frac{\partial y_2}{\partial x}\right|_{(1,t)} = Bi_m\, (1 - y_2(1,t)) \tag{6.54}$$

where n is the reaction order, $Da = k'' R_p C_{Ab}^{n-1}/D$ is the Damköhler number, Bi_m is the Biot number for mass transfer, $\delta = \vartheta_S M_S C_{Ab}/(\vartheta_A \rho_S)$ is a dimensionless number, k'' is the reaction rate, M_S is the molecular weight of reactant S, and ρ_S is the density of S. The initial conditions are:

$$y_1(x,0) = 0, \quad y_2(x,0) = 1, \quad x_c(0) = 1 \tag{6.55}$$

Finally, the interface velocity equation is of the form:

$$\frac{dx_c}{dt} = -\delta\left.\frac{\partial y_2}{\partial x}\right|_{(x_c(t),t)} \tag{6.56}$$

which, by Equation 6.53, is equivalent to

$$\frac{dx_c}{dt} = -\delta\, Da\, y_2^n(x_c(t),t) \tag{6.57}$$

As mentioned before, initially, at $t = 0$, the space region corresponding to the second phase has zero thickness. So let us consider a starting instant t_0 for the numerical simulation for which the phase change has started, and with the initial position of the interface assumed to be very close to that presented in Equation 6.55. Then we have to determine the initial instant t_0 and the initial condition $y_2(x_c, t_0)$. The initial position $x_c(0) = 1$ is substituted by :

$$x_c(t_0) = 0.999 \qquad (6.58)$$

For the linear kinetics case ($n = 1$), we can find the formula for t_0, by approximating Equation 6.56 of the interface velocity. To avoid a zero thickness region in the second phase, we consider the initial instant t_0 defined by

$$\frac{x_c - 1}{t_0} = -\delta\, Da\, y_2\,(x_c, t_0) \Leftrightarrow$$

$$\Leftrightarrow t_0 = \frac{0.001}{\delta\, Da\, y_2\,(x_c, t_0)} \qquad (6.59)$$

In order to determine the value of $y_2(x_c, t_0)$, we approximate the boundary condition (6.53) so that,

$$\frac{y_2\,(1, t_0) - y_2\,(x_c, t_0)}{1 - x_c} = Da\, y_2\,(x_c, t_0) \Leftrightarrow$$

$$\Leftrightarrow y_2\,(x_c, t_0) = \frac{y_2\,(x_c, 0)}{1 + 0.001\, Da} \Leftrightarrow$$

$$\Leftrightarrow y_2\,(x_c, t_0) = \frac{1}{1 + 0.001\, Da} \qquad (6.60)$$

For all numerical simulations, 2 finite elements in the first phase and 7 finite elements in the second phase are considered. For both meshes the starting nodes positions are equally spaced and we choose to mark 10 interior points in each finite element. Lobatto's quadrature with 33 interior quadrature points was chosen to compute all the integrals appearing in the equations of the ODE systems. Default values for the MFEM penalty functions and ODE solver tolerances were considered. The computational time required to obtain the numerical computing of solutions in the isothermal case varies between $8s$ and $100s$, and it increases with the value of the parameter Da.

The time derivative $\frac{dx_c}{dt}$ is negative for all t, thus the interface will move to the left in the space domain, and consequently we need to set a limit for that movement. In this case we set a value close to the left extreme of the domain of the first phase, $x = 10^{-3}$. When x_c reaches that value the numerical simulation ends and the amount of time taken for the moving interface to go from $x_c = 0.999$ to $x_c = 10^{-3}$ is what we call the total reaction time, t_f. This is the dimensionless time taken to react the spherical pellet completely. Figure 6.13 presents the dimensionless total reaction times as a function of the Damköhler number Da with δ as a parameter, for $Bi_m = 10^3$ and $n = 1$, using logarithmic scales. The results for the isothermal case with linear kinetics, $n = 1$, $Bi_m = 10^3$, and $\delta = 10^3$, studying the effect of the variation of

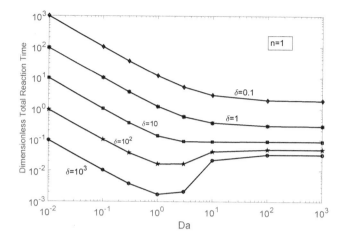

FIGURE 6.13

The shrinking core model. Isothermal case. Dimensionless total time of reaction as a function of Da with δ as a parameter for first-order kinetics (logarithmic scales).

the Damköhler number on the concentration, are described in the following three figures. The first one, Figure 6.14, shows the dimensionless concentration profiles of reactant A in phase 2 at the instants of time for which the interface positions, x_c, are $x_c = 0.1$, $x_c = 0.2$, $x_c = 0.4$, $x_c = 0.6$, $x_c = 0.8$, and $x_c = 0.9$, with $Da = 10^3$. Figure 6.14 also shows the corresponding instant of time. Then, in Figure 6.15 we can see the dimensionless rate of reaction at the interface, $Da \times y_2(x_c(t), t)$, as a function of the reduced time, normalized by the total reaction time for $Da = 0.3$, $Da = 1$, $Da = 3$, $Da = 10$, and $Da = 100$. The normalization by the dimensionless total reaction time is done to compare the rate of reaction at the interface for different values of Da. Finally, Figure 6.16 gives the dimensionless conversion at the interface, $1 - x_c^3(t)$, as function of the reduced time, normalized by the total reaction time, for $Da = 0.3$, $Da = 1$, $Da = 3$, $Da = 10$, and $Da = 100$. The values chosen for the Damköhler number, Da, illustrate three different situations. The first is when the diffusion controls the process ($Da = 10$, $Da = 100$, and $Da = 1000$); the second is when the reaction is initially kinetically controlled, but after a while the control is passed to diffusion ($Da = 3$); the last scenario is when the reaction is mainly controlled by kinetics ($Da = 0.3$ and $Da = 1$).

Nonlinear kinetics at the interface are also studied. Simulations for second-order kinetics are summarized in the plots displayed in Figure 6.17, showing the dimensionless total reaction time as a function of Da with δ as a parameter, using logarithmic scales.

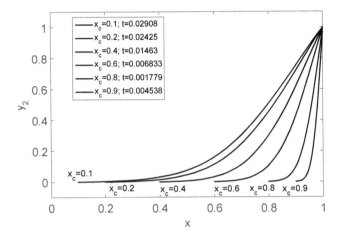

FIGURE 6.14
The shrinking core model. Isothermal case. Dimensionless concentration profiles of reactant A for the instants at which the interface x_c is at the dimensionless positions $x_c = 0.1$, $x_c = 0.2$, $x_c = 0.4$, $x_c = 0.6$, $x_c = 0.8$ and
$x_c = 0.9$.

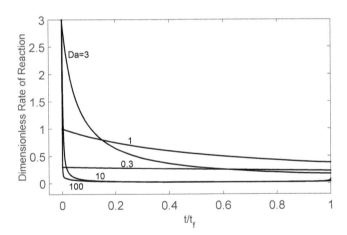

FIGURE 6.15
The shrinking core model. Isothermal case. Dimensionless rate of reaction
at the interface, $Da \times y_2(x_c(t), t)$, as function of time, normalized by total
reaction time, t_f, for $Da = 0.3$, $Da = 1$, $Da = 3$, $Da = 10$, and $Da = 100$.

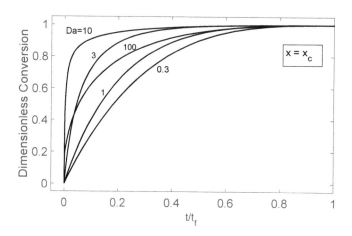

FIGURE 6.16

The shrinking core model. Isothermal case. Dimensionless conversion at the interface, $1 - x_c^3(t)$, as a function of time, normalized by total reaction time, t_f, for $Da = 0.3$, $Da = 1$, $Da = 3$, $Da = 10$, and $Da = 100$.

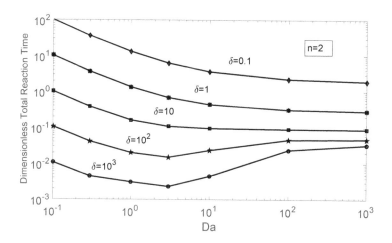

FIGURE 6.17

The shrinking core model. Isothermal case, second-order kinetics. Dimensionless total time of reaction as a function of Da with δ as a parameter (logarithmic scales).

The shrinking core model. Nonisothermal case.

Now we focus our attention on the nonisothermal case, where the reaction rate depends on the temperature at the core during the reaction. So the mathematical model for this biphasic problem should be adjusted and a new dependent variable added, namely, the dimensionless temperatures, $\theta = (T - T_b)/T_b$. We rearrange the notation of the dependent variables and denote the dimensionless temperatures at phase 1 and phase 2 y_1 and y_2, respectively. In the second phase we also consider the concentration of reactant A and this variable is denoted by y_3.

The model is now defined by mass balance equations combined with the dimensionless energy balance equations. The Damköhler number is, in this case, a function of the temperature and is given by:

$$Da = \phi e^{\frac{-\gamma}{1+\theta}} \tag{6.61}$$

where $\gamma = E/(RT_b)$ is the Arrhenius number at bulk temperature, T_b, and $\phi = A^* R_p C_{Ab}^{n-1}/D$ is a modified Damköhler number, where A^* is the preexponential factor. Thus the dependent variables should satisfy the following equations

$$Fo_1 \frac{\partial y_1}{\partial t} = \frac{\partial^2 y_1}{\partial x^2} + \frac{2}{x}\frac{\partial y_1}{\partial x}; \quad 0 < x < x_c(t), \, t > 0 \tag{6.62}$$

$$Fo_2 \frac{\partial y_2}{\partial t} = \frac{\partial^2 y_2}{\partial x^2} + \frac{2}{x}\frac{\partial y_2}{\partial x}; \quad x_c(t) < x < 1, \, t > 0 \tag{6.63}$$

$$Fo_2 \frac{\partial y_3}{\partial t} = \frac{\partial^2 y_3}{\partial x^2} + \frac{2}{x}\frac{\partial y_3}{\partial x}; \quad x_c(t) < x < 1, \, t > 0 \tag{6.64}$$

where $Fo_i = k_i/(\rho_i c_{p_i} D)$ is the Fourier number in region i, k_i is the heat conductivity in region i, and c_{p_i} is the heat capacity of i. The fixed boundary conditions are given by

$$\begin{aligned}
\frac{\partial y_1}{\partial x}\Big|_{(0,t)} &= 0 \\
\frac{\partial y_2}{\partial x}\Big|_{(1,t)} &= -Bi_h \, y_2(1,t) \\
\frac{\partial y_3}{\partial x}\Big|_{(1,t)} &= Bi_m \, (1 - y_3(1,t))
\end{aligned} \tag{6.65}$$

and, at the interface we have the boundary conditions,

$$y_1(x_c(t),t) = y_2(x_c(t),t) \tag{6.66}$$

$$\frac{\partial y_3}{\partial x}\Big|_{(x_c(t),t)} = Da \, y_3^n(x_c(t),t) \tag{6.67}$$

and

$$- \kappa \left. \frac{\partial y_1}{\partial x} \right|_{(x_c(t),t)} + \left. \frac{\partial y_2}{\partial x} \right|_{(x_c(t),t)} -$$
$$- \alpha \exp \left(\frac{-\gamma}{1 + y_1(x_c(t),t)} \right) y_3^n(x_c(t),t) = 0 \qquad (6.68)$$

where κ is the ratio between the heat conductivity of the non-porous solid, k_1, and the heat conductivity of the porous solid, k_2, $\alpha = \Delta H A^* C_{Ab}^n R_S / (T_b k_2)$ is a dimensionless number related to the heat of reaction ΔH, and Bi_h is the Biot number for the heat transfer. The imposed initial conditions are,

$$y_1(x,0) = y_2(x,0) = 0$$
$$y_3(x,0) = 1$$
$$x_c(0) = 1 \qquad (6.69)$$

The interface moves according to Equation 6.56,

$$\frac{dx_c}{dt} = -\delta \left. \frac{\partial y_3}{\partial x} \right|_{(x_c(t),t)} \qquad (6.70)$$

Now we study the effect of the parameter α on the temperature and concentration. Table 6.1 gives other model parameters.

TABLE 6.1
The shrinking core model. Model parameters,
nonisothermal case.

Bi_m	Bi_h	γ	κ	n	ϕ	δ	Fo_i
10^3	10^3	20.9	1.4	1	3.7×10^8	4.1×10^{-4}	1

Analogously to the isothermal case and to implement our MFEM strategy, no regions of zero thickness should be treated as such. Consequently, the moving interface will start its movement at a point very close to $x = 1$. For initial instant t_0, let the interface position be at

$$x_c(t_0) = 0.999 \qquad (6.71)$$

For all numerical simulations, 3 finite elements are used in the case of variable y_1, while for variables y_2 and y_3 4 finite elements are used. For all variables we choose to mark 10 interior points in each finite element. Lobatto's quadrature, with 33 interior quadrature points, was chosen to compute all the integrals appearing in the equations of the ODE systems. Default values for the MFEM penalty constants and ODE solver tolerances were considered. The computational time required for computing the numerical solutions in the non-isothermal case varies between $60s$ and $120s$. It should be remembered that a left limit is imposed on the movement of the interface. The total reaction time is the time spent for the moving interface to go from $x_c = 0.999$ to $x_c = 10^{-3}$.

The initial spatial mesh points in phase 1 are

$$X_{1,1} = 0, X_{1,2} = 0.9985, X_{1,3} = 0.99875, X_{1,4} = x_c(t_0) \qquad (6.72)$$

and, in phase 2

$$X_{i,1} = x_c(t_0), X_{i,2} = 0.99925, X_{i,3} = .99945, X_{i,4} = 0.99975, X_{i,5} = 1 \quad (6.73)$$

for $i = 2, 3$. These meshes are used to compute the initial instant t_0 of the numerical simulation. We also consider relations derived from boundary and initial conditions to determine t_0. Thus, keeping in mind what was done in the isothermal case, we arrive at the following approximations

$$\frac{x_c - 1}{t_0} = -\delta\,\phi\exp\left(\frac{-\gamma}{1+\theta_c}\right) y_3\left(x_c, t_0\right) \Leftrightarrow$$

$$\Leftrightarrow t_0 = \frac{0.001}{\delta\,\phi\exp\left(\frac{-\gamma}{1+\theta_c}\right) y_3\left(x_c, t_0\right)} \qquad (6.74)$$

and

$$\frac{y_3\left(1, t_0\right) - y_3\left(x_c, t_0\right)}{1 - x_c} = \phi\exp\left(\frac{-\gamma}{1+\theta_c}\right) y_3\left(x_c, t_0\right) \Leftrightarrow$$

$$\Leftrightarrow y_3\left(x_c, t_0\right) = \frac{1}{1 + 0.001\,\phi\exp\left(\frac{-\gamma}{1+\theta_c}\right)} \qquad (6.75)$$

On the other hand, from Equation 6.68, we can determine an initial value for θ_c, the temperature at interface x_c

$$-\kappa\frac{\theta_c - 0}{0.00025} + \frac{0 - \theta_c}{0.00025} = \alpha\exp\left(\frac{-\gamma}{1+\theta_c}\right) y_3\left(x_c, t_0\right) \Leftrightarrow$$

$$\Leftrightarrow \theta_c = \frac{-0.00025\,\alpha\exp\left(\frac{-\gamma}{1+\theta_c}\right) y_3\left(x_c, t_0\right)}{\kappa + 1} \qquad (6.76)$$

Combining the information from Equation 6.75 and Equation 6.76 we obtain the values for θ_c and $y_3\left(x_c, t_0\right)$ which are used to determine t_0 in Equation 6.74.

For the numerical simulations, we consider exothermic reactions, $\alpha < 0$, and present some figures to illustrate the system response to changes in the α parameter. Figure 6.18 gives the total reaction time versus the parameter $-\alpha$ using a logarithmic scale on the $-\alpha$ axis. It shows that the reaction time remains constant as the heat of reaction increases, but from a certain value of heat reaction it decreases, indicating a sensitivity that could trigger an almost instantaneous reaction. Thus an increase in the heat of reaction implies the decrease of the reaction time, while the interface moves faster toward the left end of the spatial domain, as Figure 6.19 illustrates.

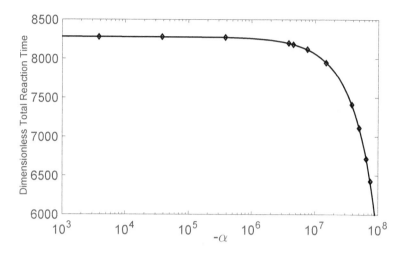

FIGURE 6.18

The shrinking core model. Nonisothermal case. Dimensionless total reaction time as a function of parameter α (logarithmic scale on $-\alpha$ axis).

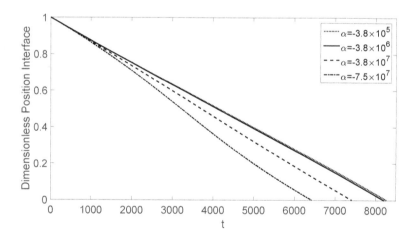

FIGURE 6.19

The shrinking core model. Nonisothermal case. Dimensionless interface position, x_c, as a function of dimensionless time for different values of α.

The dimensionless temperature at the moving interface as a function of dimensionless time can be seen in Figure 6.20. Figure 6.21 shows the dimensionless concentration at the moving interface as a function of dimensionless

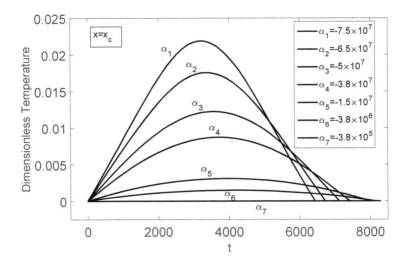

FIGURE 6.20
The shrinking core model. Nonisothermal case. Dimensionless temperature at the interface as a function of dimensionless time for different values of α.

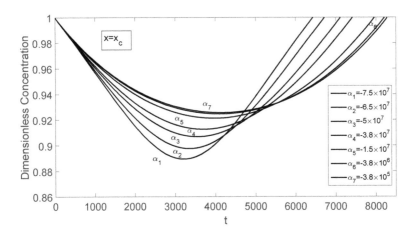

FIGURE 6.21
The shrinking core model. Nonisothermal case. Dimensionless concentration at the interface as a function of dimensionless time for different values of α.

time. Profiles were drawn considering different values of the parameter under study, $\alpha = -3.8 \times 10^5$, $\alpha = -3.8 \times 10^6$, $\alpha = -1.5 \times 10^7$, $\alpha = -3.8 \times 10^7$, $\alpha = -5.0 \times 10^7$, $\alpha = -6.5 \times 10^7$, and $\alpha = -7.5 \times 10^7$. Figure 6.22 shows the dimensionless rate of reaction, $Da \times y_3(x_c, t)$, at the interface as a function of dimensionless time, when α takes the same values.

Increasing the values of the parameter $|\alpha|$, the temperature at the interface reaches higher values (Figure 6.20). The same holds for the rate of reaction at the interface (Figure 6.22). To conclude, dimensionless temperature profiles and dimensionless concentration profiles as a function of the space variable are exhibited in Figure 6.23 and Figure 6.24, respectively, for the simulation with $\alpha = -7.5 \times 10^7$. The profiles are taken at different instants of time, corresponding to the position of the moving boundary at $x_c = 0$ to $x_c = 1$ with step 0.1. The dashed line in both figures represents the temperature and concentration at the moving boundary. The results obtained with our MFEM based technique are in good agreement with those presented by Carabin et al. [17]. These authors used finite elements in space and finite differences in time and also Newton–Raphson iterations to handle the nonlinearity in the boundary conditions. The strategy developed for initializing the numerical simulation, combined with the MFEM, is straightforward and provides results with a degree of accuracy.

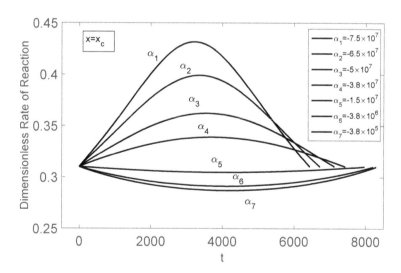

FIGURE 6.22

The shrinking core model. Nonisothermal case. Dimensionless rate of reaction $(Da \times y_3(x_c, t))$ at the interface as a function of dimensionless time for different values of α.

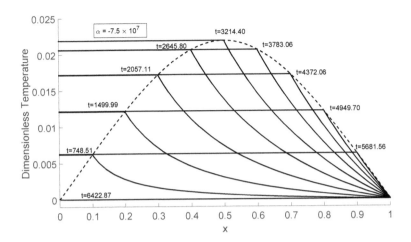

FIGURE 6.23
The shrinking core model. Nonisothermal case. Dimensionless temperature as a function of the dimensionless space variable for $\alpha = -7.5 \times 10^7$ (solid) and temperature at the moving boundary (dashed).

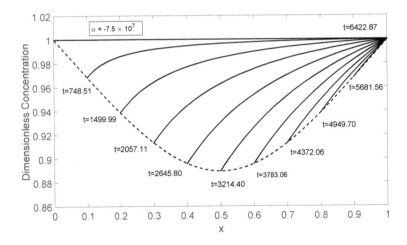

FIGURE 6.24
The shrinking core model. Nonisothermal case. Dimensionless concentration as a function of the dimensionless space variable for $\alpha = -7.5 \times 10^7$ (solid) and concentration at the moving boundary (dashed).

6.4   Causticizing Reaction

In this section, we solve a causticizing reaction model which is an important step of the Kraft paper process of pulp production. It is assumed that the reaction is heterogeneous and that the diffusion controls all the process, forcing the reaction to occur in a solid-liquid moving interface, [18]. Lime mud particles react with sodium carbonate and the result is the production of sodium hydroxide and calcium carbonate. What we are interested in is the diffusion of the ions OH^- and CO_3^{2-} as in Figure 6.25.

This is a biphasic problem, with phase 1 corresponding to the solid phase and phase 2 to the liquid phase. In the mathematical model, the dependent variables will be the ion concentrations, expressed in mol/l. So y_1 and y_3 are the OH^- concentrations in the first phase and second phase, respectively. Analogously, y_2 and y_4 are the CO_3^{2-} concentrations in the first phase and second phase, respectively. The spatial independent variable x is the normalized radial position via the radius of the sphere, R. The normalized position of the interface is $x_c(t)$, t being the time variable expressed in seconds. Besides, the others parameters of the model are the diffusivity of ions, De; the equilibrium constant, K_{eq}; the mass transfer coefficient, k_L; the porosity of particles, ϵ; and the initial concentration of $Ca(OH)_2$ denoted by $C_{Ca(OH)_2}$.

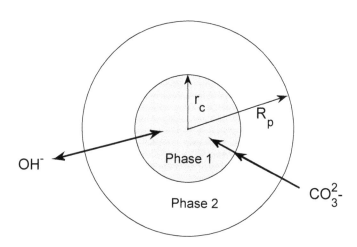

FIGURE 6.25
Causticizing reaction. Biphasic problem of causticizing reaction: Ion diffusion.

The partial differential equations that describe the process are the following,

$$\frac{\partial y_1}{\partial t}(x,t) = \frac{De}{\epsilon R^2}\left(\frac{\partial^2 y_1}{\partial x^2} + \frac{2}{x}\frac{\partial y_1}{\partial x}\right) \tag{6.77}$$

$$\frac{\partial y_2}{\partial t}(x,t) = \frac{De}{\epsilon R^2}\left(\frac{\partial^2 y_2}{\partial x^2} + \frac{2}{x}\frac{\partial y_2}{\partial x}\right) \tag{6.78}$$

in phase 1, $0 < x < x_c(t)$, $t > 0$ and

$$\frac{\partial y_3}{\partial t}(x,t) = \frac{De}{\epsilon R^2}\left(\frac{\partial^2 y_3}{\partial x^2} + \frac{2}{x}\frac{\partial y_3}{\partial x}\right) \tag{6.79}$$

$$\frac{\partial y_4}{\partial t}(x,t) = \frac{De}{\epsilon R^2}\left(\frac{\partial^2 y_4}{\partial x^2} + \frac{2}{x}\frac{\partial y_4}{\partial x}\right) \tag{6.80}$$

in phase 2, $x_c(t) < x < 1$, $t > 0$. The dependent variables are subjected to the following boundary conditions,

$$(y_1(x_c(t),t))^2 = K_{eq}\, y_2(x_c(t),t) \tag{6.81}$$

$$y_1(x_c(t),t) = y_3(x_c(t),t) \tag{6.82}$$

$$y_2(x_c(t),t) = y_4(x_c(t),t) \tag{6.83}$$

and

$$\left.\frac{\partial y_3}{\partial x}\right|_{(x_c(t),t)} - \left.\frac{\partial y_1}{\partial x}\right|_{(x_c(t),t)} - 2\left.\frac{\partial y_2}{\partial x}\right|_{(x_c(t),t)} + 2\left.\frac{\partial y_4}{\partial x}\right|_{(x_c(t),t)} = 0 \tag{6.84}$$

for the moving boundary, and

$$\frac{\partial y_i}{\partial x}(0,t) = 0, \quad i = 1,2 \tag{6.85}$$

$$\left.\frac{\partial y_i}{\partial x}\right|_{(1,t)} = \frac{R\,k_L}{De}\left(\psi_i(t) - y_i(1,t)\right), \quad i = 3,4 \tag{6.86}$$

for fixed boundaries, where ψ_3 and ψ_4 represent the concentrations of the ions OH^- and CO_3^{2-} around the particles, respectively. Functions ψ_3 and ψ_4 are defined by

$$\psi_3(t) = \begin{cases} 2.144 \text{ mol/l}; & 600 \le t \le 2400 \\ 2.254 \text{ mol/l}; & 2400 < t \le 4200 \\ 2.279 \text{ mol/l}; & 4200 < t \le 6000 \end{cases} \tag{6.87}$$

$$\psi_4(t) = \begin{cases} 0.228 \text{ mol/l}; & 600 \le t \le 2400 \\ 0.223 \text{ mol/l}; & 2400 < t \le 4200 \\ 0.211 \text{ mol/l}; & 4200 < t \le 6000 \end{cases} \tag{6.88}$$

Initial conditions are

$$y_1(x,t_0) = y_2(x,t_0) = 0\,\text{mol/l} \tag{6.89}$$

$$y_3(x,t_0) = 2.025\,\text{mol/l} \tag{6.90}$$

$$y_4(x,t_0) = 0.337\,\text{mol/l} \tag{6.91}$$

$$x_c(t_0) = 0.8$$

The reaction front velocity is given by

$$\frac{dx_c}{dt} = -\frac{De}{R^2\, C_{Ca(OH)_2}} \left(\frac{\partial y_4}{\partial x}\bigg|_{(x_c(t),t)} - \frac{\partial y_2}{\partial x}\bigg|_{(x_c(t),t)} \right) \tag{6.92}$$

The model parameters, taken from [18], are given in Table 6.2.

TABLE 6.2

Causticizing reaction. Model parameters.

De	K_{eq}	k_L	ϵ	R	$C_{Ca(OH)_2}$
7.2×10^{-12} m^2/s	50 mol/l	5×10^{-4} m/s	0.1	$\sqrt{10^{-9}}$ m	27 mol/l

In all the numerical simulations we use 4 finite elements concerning each one of the spatial meshes. The initial normalized position of the moving interface is $x_c = 0.8$. We chose as initial mesh points,

$$X_{i,1} = 0, X_{i,2} = 0.5, X_{i,3} = 0.6, X_{i,4} = 0.7, X_{i,5} = 0.8 \tag{6.93}$$

where $i = 1, 2$, i.e., for the mesh associated with phase 1, and

$$X_{i,1} = 0.8, X_{i,2} = 0.85, X_{i,3} = 0.9, X_{i,4} = 0.999, X_{i,5} = 1 \tag{6.94}$$

where $i = 3, 4$, i.e., for the mesh associated with phase 2. The number of interior points per finite element is 2 in the first phase and 3 in the second phase. Figure 6.26 shows the node movement for meshes associated with hydroxide ions, OH^- (left) and for carbonate ions, CO_3^{2-} (right), along time, and the good distribution of those nodes in both phases is verified. Solid lines

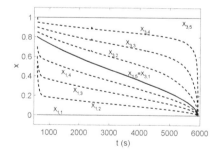
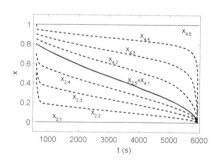

FIGURE 6.26

Causticizing reaction. Node movement for meshes associated to hydroxide ions, OH^- (left) and to carbonate ions, CO_3^{2-} (right), as a function of time (s).

correspond to the moving interface and to the fixed boundaries $x = 0$ and $x = 1$.

In this work, we consider $t_0 = 600s$ for the starting instant of time, because the preceding step of causticizing, the slaking, takes approximately 10 minutes [18]. In order to define the starting solutions at the mesh points, we consider numerical approximations to the boundary conditions defined by Equation 6.85 and Equation 6.86 together with the initial conditions established by Equation 6.89, Equation 6.90, and Equation 6.91.

Concerning the MFEM parameters, default values are used and the computational time required for the numerical computing of solutions is approximately $9s$.

Notice that with the MFEM formulation, all the phases must have non-zero thickness. Therefore we need to mark a limit for the movement of the interface, so that in this problem the first phase always has positive length. The integration is performed until the position of the interface is close enough, $x = 10^{-3}$, to the left end of the physical spatial domain. This position of the moving interface is achieved at time instant $t = 5935s$. In Figure 6.27 is highlighted the evolution with time of the position of the moving interface. It is almost linear except for values close to $x = 0$, where the reaction is faster due to the role of diffusion.

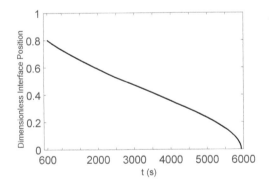

FIGURE 6.27
Causticizing reaction. Normalized moving interface position, as a function of time (s).

The concentrations at the fixed boundaries correspond to three different reactors with a total retention time close to $5400s$, which agree with the industrial data, [18]. The retention time in each reactor is enough to reach smooth and stable solution profiles. Figure 6.28 gives the histories of concentration at $x = 1$ as a function of time. Figure 6.28 (top) shows the concentration of hydroxide ions OH^- at $x = 1$ as a function of time while Figure 6.28 (bottom) presents the concentration of carbonate ions CO_3^{2-} at $x = 1$ as a function of time.

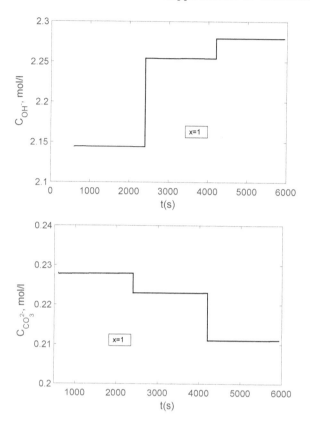

FIGURE 6.28

Histories of concentration at $x = 1$ as a function of time for hydroxide ions, OH^- (top) and for carbonate ions, CO_3^{2-} (bottom).

Finally concentration profiles at different time instants are presented. Figure 6.29 exhibits the concentration profiles at the time instants $t = 604s$, $t = 2396s$, $t = 2404s$, $t = 4196s$, $t = 4204s$, $t = 5800s$, and $t = 5935s$, of hydroxide ions OH^- as a function of reduced space variable x. Figure 6.30 presents the concentration profiles of carbonate ions CO_3^{2-} at the same time instants, as a function of reduced space variable x.

The numerical results obtained are in good agreement with the ones presented by Duarte and Portugal [18], who used a moving finite element method with Hermite polynomials and DASOLV, an integrator for large, sparse systems of differential and algebraic equations (DAEs) [19].

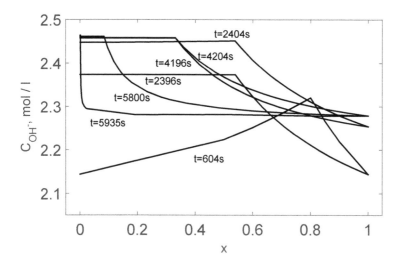

FIGURE 6.29
Concentration profiles of hydroxide ions, OH^- at the instants $t = 604s$, $t = 2396s$, $t = 2404s$, $t = 4196s$, $t = 4204s$, $t = 5800s$, and $t = 5935s$ as a function of reduced space variable x.

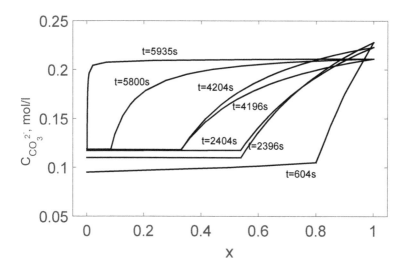

FIGURE 6.30
Concentration profiles of carbonate ions, CO_3^{2-}, at the instants $t = 604s$, $t = 2396s$, $t = 2404s$, $t = 4196s$, $t = 4204s$, $t = 5800s$, and $t = 5935s$ as a function of reduced space variable x.

6.5   Hydride Electrodes

Here we present a shrinking core model for the galvanostatic discharge of a metal hydride spherical particle [20], used to make negative electrodes. There is a metal-electrolyte moving interface and hydrogen diffusion within the bulk of the particle. The discharge process is shown schematically in Figure 6.31 with a plot of the hydrogen concentration profile and distinction between the 2 phases of the problem. The first phase is called phase β, while the second is phase α. r_0 is the radius of the spherical particle and r_α is the position of the interface. We denote by $C_{\beta,\alpha}$ the initial hydrogen concentration inside the particle, C_s the surface concentration and $C_{\alpha,\beta}$ the hydrogen concentration at the interface and C_0 represents the ratio $C_0 = \frac{C_{\beta,\alpha}}{C_{\alpha,\beta}}$. The consumption of adsorbed hydrogen promotes the formation and increase of phase α, according to the reaction $H_{ads} + OH^- = H_2 0 + e^-$. There is diffusion of hydrogen atoms from phase β to the surface through phase α.

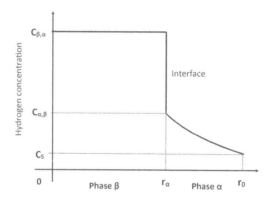

FIGURE 6.31
Hydride electrodes. Biphasic problem: hydrogen concentration profile in a metal hydride particle.

While in phase β the hydrogen concentration remains constant, in phase α the concentration satisfies the equation,

$$\frac{\partial C}{\partial \tau} = \frac{D_\alpha}{r^2} \frac{\partial}{\partial r} \left(r^2 \frac{\partial C}{\partial r} \right) \qquad (6.95)$$

where C is the concentration expressed in mol/cm^3, r is the radial position expressed in cm, and τ is the time in seconds and where D_α is the diffusion coefficient in phase α. At the interface it must be considered a mass conservation law,

$$(C_0 - 1) \frac{dr_\alpha}{d\tau} = D_\alpha \left. \frac{\partial C}{\partial r} \right|_{r=r_\alpha} = -\frac{I \, r_0^3 \, \rho}{3 \, F \, r_\alpha^2} \tag{6.96}$$

where I is the applied current, F is the Faraday constant, and ρ is the mass density of the particle.

In the next steps we normalize the space variable and make dimensionless the concentration variables. For that purpose consider the following transformation of variables:

$$x = \frac{r}{r_0}; \quad x_\alpha = \frac{r_\alpha}{r_0}; \quad y_1 = \frac{C}{C_{\alpha,\beta}}; \quad y_2 = \frac{C}{C_{\alpha,\beta}}; \quad t = \frac{D_\alpha \tau}{r_0^2} \tag{6.97}$$

Now x is the reduced space variable, x_α is the reduced moving interface position and t is the reduced time variable, while y_1 and y_2 are the dimensionless hydrogen concentrations at phase β and α, respectively. Replacing the variables in Equations 6.96 by those defined by Equation 6.97, we obtain the following conditions:

$$\left. \frac{\partial y_2}{\partial x} \right|_{x=x_\alpha} = -\frac{I \, r_0^4 \, \rho}{3 \, F \, r_\alpha^2 \, D_\alpha \, C_{\alpha,\beta}} \tag{6.98}$$

$$\frac{dx_\alpha}{dt} = \frac{1}{C_0 - 1} \left. \frac{\partial y_2}{\partial x} \right|_{x=x_\alpha} \tag{6.99}$$

The model parameters are δ (dimensionless current density) and κ defined by:

$$\delta = \frac{I \, r_0^4 \, \rho}{3 \, F \, r_\alpha^2 \, D_\alpha \, C_{\alpha,\beta}} \quad ; \quad \kappa = \frac{1}{C_0 - 1} \tag{6.100}$$

Thus, the normalized model is defined by

$$\frac{\partial y_1}{\partial t} = 0 \quad ; \quad 0 < x < x_\alpha(t) \, , \, t > 0 \tag{6.101}$$

$$\frac{\partial y_2}{\partial t} = \frac{\partial^2 y_2}{\partial x^2} + \frac{2}{x} \frac{\partial y_2}{\partial x} \quad ; \quad x_\alpha(t) < x < 1 \, , \, t > 0 \tag{6.102}$$

and the dependent variables must satisfy the following boundary conditions:

$$y_1(0, t) = 1 + \frac{1}{\kappa} \tag{6.103}$$

$$y_1(x_\alpha(t), t) = 1 + \frac{1}{\kappa} \tag{6.104}$$

$$y_2(x_\alpha(t), t) = 1 \tag{6.105}$$

$$\left. \frac{\partial y_2}{\partial x} \right|_{x=1} = -\delta \tag{6.106}$$

and the initial conditions are

$$y_1(x,0) = 1 + \frac{1}{\kappa} \tag{6.107}$$

$$y_2(x,0) = 1 + \frac{1}{\kappa} \tag{6.108}$$

$$x_\alpha(0) = 1 \tag{6.109}$$

Finally, the equation for the moving interface velocity is

$$\frac{dx_\alpha}{dt} = \kappa \left. \frac{\partial y_2}{\partial x} \right|_{x_\alpha(t)} \tag{6.110}$$

In the numerical simulations we consider $C_{\beta,\alpha} = 0.86$ and $C_{\alpha,\beta} = 0.1$, giving $\kappa = 0.1316$ [21], and different values of δ parameter.

One more time, for our MFEM formulation it is not possible to have zero-thickness regions. Therefore we consider an initial instant t_0, so that the interface is very close to the right extreme of the spatial domain. In this case we chose:

$$x_\alpha(t_0) = 0.9998 \tag{6.111}$$

The values of t_0 and $y_2(1, t_0)$ depend on the parameter δ and are calculated based on approximations of the boundary conditions and initial conditions.

For all numerical simulations we consider 2 finite elements in the first phase and 4 finite elements in the second phase. The number of interior points in each finite element is 2 for the first phase and 5 for the second. With respect to the starting grids at $t = t_0$, we take

$$X_{1,1} = 0, X_{1,2} = 0.5, X_{1,3} = 0.9998 \tag{6.112}$$

for the mesh associated to the first phase, phase β, and

$$X_{2,1} = 0.9998, X_{2,2} = 0.99985, X_{2,3} = 0.9999,$$
$$X_{2,4} = 0.99995, X_{2,5} = 1 \tag{6.113}$$

for the mesh associated to the second phase, phase α. According to the initial conditions, the initial values at time instant t_0 for dimensionless hydrogen concentration at node positions are $1 + \frac{1}{\kappa}$ for the mesh nodes on phase β and α, except for $y_2(X_{2,5}(t_0), t_0)$. This value is calculated according to the discretization of the boundary conditions by using nodes $X_{2,4}$ and $X_{2,5}$ in Equation 6.106

$$y_2(X_{2,5}(t_0), t_0) \quad = \quad y_2(X_{2,4}(t_0), t_0) - (X_{2,5}(t_0) - X_{2,4}(t_0))\delta \tag{6.114}$$

and discretizing Equation 6.110 given the interface velocity by using the position of interface at time instants $t = 0$ and $t = t_0$, $x_\alpha(0)$ and $x_\alpha(t_0)$, and the value of y_2 at $X_{2,4}(t_0)$ and $X_{2,5}(t_0)$.

$$\frac{x_\alpha(t_0) - x_\alpha(0)}{t_0 - 0} = k \frac{y_2(X_{2,4}(t_0), t_0) - y_2(X_{2,5}(t_0), t_0)}{x_\alpha(t_0) - x_\alpha(0)} \qquad (6.115)$$

Thus

$$y_2(X_{2,5}(t_0), t_0) = 1 - (1 - 0.99995)\delta \qquad (6.116)$$

$$t_0 = \frac{(0.9998 - 1)^2}{k(1 - y_2(X_{2,5}(t_0), t_0))} \qquad (6.117)$$

Default values are used for numerical simulations and the computational time spent in the computation is approximately $3s$. Once again, to prevent a zero-thickness first phase we put a limit to the movement of the interface, $x = 10^{-4}$.

The results of numerical simulations are shown in Figure 6.32 and Figure 6.33. Figure 6.32 presents the dimensionless interface positions as functions of dimensionless time for different values of current density: $\delta = 1$, $\delta = 2$, $\delta = 5$ and $\delta = 50$. Our numerical solutions are compared with pseudo-steady-state (PSS) solution, which in phase α can be obtained by putting equal to zero the term $\frac{\partial y_2}{\partial t}$ of Equation 6.102:

$$y_{2,PSS} = 1 + \delta \left(\frac{1}{x} - \frac{1}{x_{\alpha,PSS}} \right) \qquad (6.118)$$

The position of the interface may be predicted by using Equation 6.118 in Equation 6.110. In the pseudo-steady-state we have

$$x_{\alpha,PSS} = (1 - 3\kappa \delta t)^{\frac{1}{3}} \qquad (6.119)$$

Our numerical solution and the PSS solution are similar for $\delta = 1$ and $t \leq 1$, but for $\delta \geq 2$, there is a notorious difference, with the hydrogen concentration given by PSS always smaller than the ones obtained by our MFEM framework. This is due to the fact that when we use the PSS approach, we are neglecting the initial hydrogen concentration value in phase α, given by Equation 6.108.

Figure 6.33 shows the dimensionless hydrogen concentration in phase α, with dimensionless current density $\delta = 1$ and $\kappa = 0.1316$, for several instants: $t = 0.5$, $t = 1.0$, $t = 1.5$, $t = 2.0$, and $t = 2.5$ as a function of normalized space variable.

In [20] the authors present another strategy for the problem. For the hydrogen concentration in phase α, the authors propose the following decomposition:

$$y_2(x, t) = u(x, t) + w(x) \qquad (6.120)$$

where w is the term corresponding to PSS solution and u takes into account the initial value of the hydrogen concentration. Our numerical results are in good agreement and the conclusions of the benchmark with PSS solution are similar to those presented in [20].

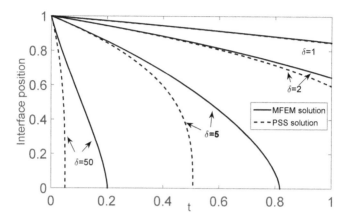

FIGURE 6.32
Hydride electrodes. Dimensionless interface position as a function of dimensionless time for values of current density, $\delta = 1$, $\delta = 2$, $\delta = 5$, and $\delta = 50$ and benchmark with pseudo-steady-state solution.

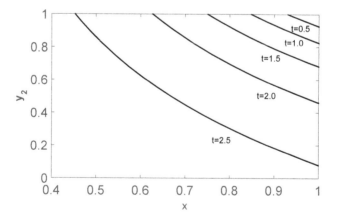

FIGURE 6.33
Hydride electrodes. Dimensionless hydrogen concentration in phase α as a function of normalized space variable, with $\delta = 1$, for reduced times $t = 0.5$, $t = 1.0$, $t = 1.5$, $t = 2.0$, and $t = 2.5$.

6.6 Swelling of Wool Grease

The swelling of wool grease in a wool scour is modeled by Barry and Caunce [22] as a Stefan problem with two moving boundaries, considering a three-phase problem, where those authors emphasize the relative limitations of using the pseudo-steady-state assumption. Here we apply our MFEM numerical scheme for both linear and nonlinear diffusivity, varying the models parameters. The spatial domain is decomposed into three regions, the non-swollen zone, the swollen zone, and the water zone. Without loss of generality, the initial positions of the boundaries are both taken to be $x = 0$, where x is the normalized spatial variable by a physical scale ξ_0 and the normalized spatial domain is $[-L, L]$ with L large enough to taking into account the boundaries trajectories. Also, the reduced time variable, t, is the time variable reduced by the typical time scale ξ_0^2/D_1, which is related to the typical length scale and to diffusion.

The normalized dependent variable in each phase, y_i $(i = 1, 2, 3)$, is the normalized moisture content, i.e., the normalized volume fraction of water. So in the first phase y_1 is null and in third phase y_3 is equal to 1. The boundary between nonswollen and swollen phases is denoted by $s(t)$, while the boundary between the swollen and water phases is denoted by $r(t)$ as Figure 6.34 shows.

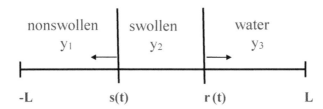

FIGURE 6.34
Swelling of wool grease. Non-swollen, swollen, and water phases.

The dimensionless partial differential equations describing the problem are,

$$\frac{\partial y_1}{\partial t} = 0, \qquad -L < x < s(t)\,,\, t > 0 \qquad (6.121)$$

$$\frac{\partial y_2}{\partial t} = \frac{\partial}{\partial x}\left(D(y_2)\frac{\partial y_2}{\partial x}\right), \qquad s(t) < x < r(t)\,,\, t > 0 \qquad (6.122)$$

$$\frac{\partial y_3}{\partial t} = 0, \qquad r(t) < x < L\,,\, t > 0 \qquad (6.123)$$

where the reduced diffusion is denoted by D and may depend on the y_2 variable. Concerning the required boundary conditions, we have,

$$y_1(-L,t) = 0, \qquad y_1(s(t),t) = 0 \qquad\qquad (6.124)$$

$$y_2(s(t),t) = 0, \qquad y_2(r(t),t) = 1 \qquad\qquad (6.125)$$

$$y_3(r(t),t) = 1, \qquad y_3(L,t) = 1 \qquad\qquad (6.126)$$

and the initial conditions are set to be:

$$y_1(x,0) = y_2(x,0) = 0, \qquad y_3(x,0) = 1 \qquad\qquad (6.127)$$

$$s(0) = 0, \qquad r(0) = 0 \qquad\qquad (6.128)$$

The positive parameters b and a determine the velocities of the two interfaces and the normalized moving boundaries equations have the form,

$$\frac{ds}{dt} = -b\left.\frac{\partial y_2}{\partial x}\right|_{(s(t),t)} \qquad\qquad (6.129)$$

$$\frac{dr}{dt} = a\left.\frac{\partial y_2}{\partial x}\right|_{(r(t),t)} \qquad\qquad (6.130)$$

To define the spatial domain, the value of L is set to $L = 5$. Choosing $t_0 = 10^{-6}$ to start the numerical simulation, the starting positions of the two boundaries and the starting solution in the second phase are estimated using the pseudo-steady-state assumption presented by Barry [22]. For the numerical simulations we consider two types of diffusion: linear and nonlinear.

To illustrate the linear case, the diffusion is chosen to be $D(y_2) = 1$. Defining the new normalized variable by

$$z = \frac{x - s(t_0)}{r(t_0) - s(t_0)} \qquad\qquad (6.131)$$

and the quantities

$$c_R = \sqrt{\frac{2\,a^2}{a+b}} \qquad\qquad (6.132)$$

$$c_S = \sqrt{\frac{2\,b^2}{a+b}} \qquad\qquad (6.133)$$

then, assuming the pseudo-steady-state, the starting positions of the boundaries and the starting solution between boundaries are given by:

$$r(t_0) = c_R\sqrt{t_0} \qquad\qquad (6.134)$$
$$s(t_0) = -c_S\sqrt{t_0} \qquad\qquad (6.135)$$
$$y_2(x,t_0) = z \qquad\qquad (6.136)$$

For the nonlinear case, the diffusion is chosen to be a standard exponential, $D(y_2) = \exp(\beta y_2 - \beta)$. Now $c_S = \gamma c_R$, where $\gamma = \frac{b\,D(1)}{a\,D(0)}$, and c_R are defined by

$$c_R = \sqrt{\frac{2a(1 - e^{-\beta})}{\beta(1 + \gamma)}} \qquad (6.137)$$

Thus, the starting positions of the boundaries and the starting solutions between boundaries are

$$r(t_0) = c_R\sqrt{t_0} \qquad (6.138)$$
$$s(t_0) = -c_S\sqrt{t_0} \qquad (6.139)$$
$$y_2 = 1 + \frac{1}{\beta}\ln\left[(1 - e^{-\beta})z + e^{-\beta}\right] \qquad (6.140)$$

For the numerical simulations, 4 finite elements in each phase with 2 interior points in each finite element were used. To compute the integrals appearing in the equations of the ODE systems we use Lobatto's quadrature with 5 interior quadrature points. The CPU time is less than 17 seconds.

If the diffusion is linear we know the analytical solution of the problem. In the other case for which the analytical solution is unknown, we can use the pseudo-steady-state solution to compare with the MFEM numerical solution. We obtain the pseudo-steady-state solution by assuming that the interfaces move slowly enough that the swollen phase has time to reach steady state before the interfaces move.

Beginning with the scenario of a linear diffusion, $D = 1$, we can see in Figure 6.35 and Figure 6.36 that the numerical solution is identical to the exact one when the parameters a and b take values 1 or 10, respectively. We can also see that for larger values of the parameters a and b the pseudo-steady-state approximation differs from both numerical and exact solutions.

From Figure 6.37, the relation between parameters a and b and the interface velocities is clear. For $a = b$ the positions of the interfaces are symmetric. In Figure 6.38 we observe that for $a \neq b$ the symmetry of the interface movement is lost. Fixing one of the parameters and decreasing the other, the interface corresponding to the first parameter moves even faster. In terms of behavior, the interface movement it is proportional to $\sqrt{t}$.

Now we present the scenario of a nonlinear diffusion case. Taking $D(y_2) = e^{2y_2 - 2}$ then the diffusion coefficient $D(y_2)$ is smaller than 1, and will influence the interface movement. Figure 6.39 and Figure 6.40 show the dimensionless moisture in the swollen phase as a function of the normalized space variable z at time instant $t = 2$. For smaller values of parameters a and b the MFEM solution is similar to the pseudo-steady-state solution, as can be seen in Figure 6.39. Despite a diffusion smaller than 1, when $a = b = 0.1$ the solution is driven by the diffusion and because of this, the solution is concave. Taking $a = b$ and increasing these parameters' values, we can conclude from Figure 6.40 that

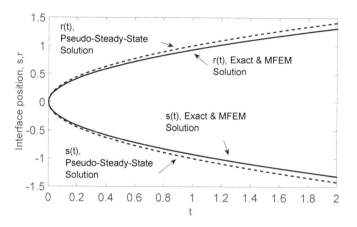

FIGURE 6.35
Swelling of wool grease. Interface positions, nonswollen/swollen (s) and swollen/water (r) as a function of reduced time for linear case and $a = b = 1$. Comparison with the exact and the pseudo-steady-state solution.

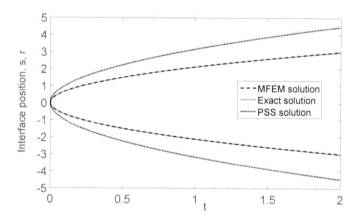

FIGURE 6.36
Swelling of wool grease. Interface positions, nonswollen/swollen (s) and swollen/water (r) as a function of reduced time for linear case and $a = b = 10$. Comparison with the exact and the pseudo-steady-state solution.

there is no concave solution and in terms of behavior it becomes closer to the scenario of linear diffusion, with the numerical solution being convex near the interface $s(t)$. On the other hand, larger values of parameter a imply a larger neighborhood of interface $r(t)$ with values of moisture closer to one.

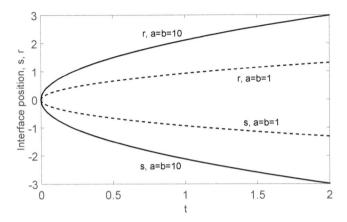

FIGURE 6.37
Swelling of wool grease. Interface positions, nonswollen/swollen (s) and swollen/water (r) as a function of reduced time for linear case and $a = b = 1$ (dashed) and $a = b = 10$ (solid).

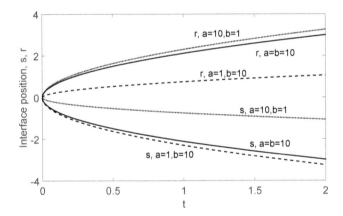

FIGURE 6.38
Swelling of wool grease. Interface positions, nonswollen/swollen (s) and swollen/water (r) as a function of reduced time for linear case and $a = b = 10$ (solid) $a = 10$, $b = 1$ (dotted) and $a = 1$, $b = 10$ (dashed).

This fact emphasizes the role of the interface velocities on the solution when the diffusion is smaller.

The present method based on MFEM when applied to this multiphase problem had to be rethought due to the issue of a starting domain of zero

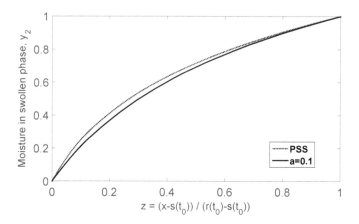

FIGURE 6.39
Swelling of wool grease. Dimensionless moisture in the swollen phase as a function of normalized space variable z at time instant $t = 2$ with $a = b = 0.1$. MFEM (solid) vs PSS (dotted) solution.

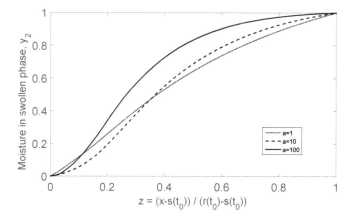

FIGURE 6.40
Swelling of wool grease. Dimensionless moisture in the swollen phase as a function of normalized space variable z at time instant $t = 2$ for different values of a, $b = a$.

thickness. The pseudo-steady-state solution has obvious limitations, but we used it only to obtain an accurate numerical initialization in a spatial domain with length close to zero. MFEM numerical results are accurate and in agreement with those presented in [22], where besides the pseudo-steady-state solutions, the authors combined analytical and numerical procedures based on finite differences method.

6.7 Frank Spheres Problem

In this section, we consider the problem proposed by Frank [23] in 1950, studying the radially symmetric phase growth controlled by diffusion. This problem has been used by several authors [24, 25, 26, 27, 28, 29] to test the accuracy of the numerical solution obtained with their algorithms. Consider the solidification problem, in the two-dimensional space domain, modeled by the dimensionless diffusion equation,

$$\frac{\partial u}{\partial t} = \nabla^2 u \tag{6.141}$$

where $u = u(\boldsymbol{x}, t)$ is the dimensionless temperature in the liquid phase. Initially, the liquid occupies the entire plane except the $(0,0)$ point, so the moving boundary expands from the origin, which constitutes the starting solid phase. Let r be the radial position of $\boldsymbol{x} = (x,y)$, $r = \left(x^2 + y^2\right)^{1/2}$. Assuming the boundary conditions

$$u\big|_{\Gamma_M} = 0, \quad u_{\Gamma_F} = u_B < 0, \quad t > 0 \tag{6.142}$$

$$\frac{\partial u}{\partial \mathbf{n}}\bigg|_{\Gamma_M} = -\dot{\boldsymbol{x}} \cdot \mathbf{n}, \quad t > 0 \tag{6.143}$$

where u_B is the value of u at the fixed boundary, in the infinity and $\mathbf{n}$ is the outward normal vector to the boundary of the relevant spatial domain. To describe the solution of this problem let s be the reduced radial position, $s = r\sqrt{t}$, and assume that u_∞ denotes the value of u when s goes to infinity. The exact solution to this Stefan problem with a single phase is

$$u(r,t) = \begin{cases} u_\infty \left(1 - \dfrac{E_1(s^2/4)}{E_1(S^2/4)}\right), & s \geq S, \\ 0, & s < S, \end{cases} \tag{6.144}$$

where E_1 is the integral exponential function defined by $E_1(z) = \int_z^\infty t^{-1} e^{-t} dt$. The exact radius of the moving interface at time instant t is $R(t) = S\sqrt{t}$. The value of the parameter S is given by the equation

$$u_\infty = -\frac{S^2}{4} e^{\frac{S^2}{4}} E_1\left(\frac{S^2}{4}\right). \tag{6.145}$$

According to [24, 25, 27], we consider $u_\infty = -0.5$. From Equation 6.145 the value of S is $S \simeq 1.5621239$.

Since at $t = 0$ the solid phase has null measure, it is necessary to make an approximation of the model to start the MFEM. Looking at the geometry of the spatial domain occupied by the solid at $t > 0$, we start the simulations at a time $t_0 > 0$ and we shall consider the limited spatial domain,

$$D(t_0) = \left\{ \boldsymbol{x} = (x, y) : R(t_0) \leq \sqrt{x_1^2 + x_2^2} \leq R_F \right\}$$

The circumference of radius $R(t_0)$ defines the starting solid–liquid interface. The numerical results were obtained using a finite element mesh which approximates, at each instant, the annulus $D(t)$, contained in $D(t_0)$, $t > t_0$. Figure 6.41 illustrates the spatial mesh defining the initial computational domain $\widetilde{D}(t_0)$, with $R(t_0) = 0.5$ and $R_F = 2$. The initial value t_0 was chosen so that the radius $R(t_0)$ has the value 0.5. The exact solution is used to define initial and boundary conditions. Note that the nodal edges in the inner boundary $\widetilde{D}(t_0)$ do not belong to $D(t_0)$. So we are introducing an initial error by including in the liquid region points that belong to the region occupied by the solid.

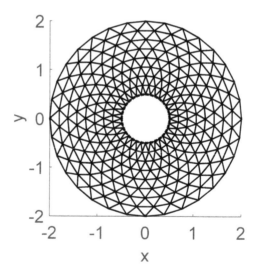

FIGURE 6.41
Frank spheres problem. Triangular decomposition of the computational domain $\widetilde{D}(t_0)$, with $R(t_0) = 0.5$ and $R_F = 2$.

This problem is described in an irregular 2D spatial domain and has only one active phase. For the application of the MFEM we consider a spatial mesh discretizing the active phase. Furthermore it is necessary to discretize the boundary condition given by Equation 6.143 to include it in the ODE system that is obtained from minimization problem. Besides 6.8 another ODE associated with $\boldsymbol{X}_{s_j} = \big((X_1)_{s_j}, (X_2)_{s_j}\big)$ must be added, requiring that the moving boundary nodes have a radial motion. Thus, the vertex coordinates must also satisfy the equation

$$(X_1)_{s_j}(\dot{X}_2)_{s_j} - (X_2)_{s_j}(\dot{X}_1)_{s_j} = 0 \qquad (6.146)$$

In this example, were considered quadratic approximations in each of the 512 initial spatial finite element mesh, shown in Figure 6.41. This discretization leads to an ODE system with 1472 equations, in the independent variable time, enabling the determination of the approximate solution of the problem described earlier in this section. The coefficient matrix of this system has the configuration shown in Figure 6.42 and has a higher degree of sparsity, greater than 98.4%, with 33,216 non-zero positions. It should be noted that the ODE integrator is provided with the pattern of nonzero elements of the matrix, allowing a substantial reduction in calculation time. It should be pointed out that the structure of the matrix represented in Figure 6.42 is obtained due to the strategy used in ordering the components of the vector of all unknown variables. The amplitudes at the interpolation points and the coordinates of the nodal positions are interspersed.

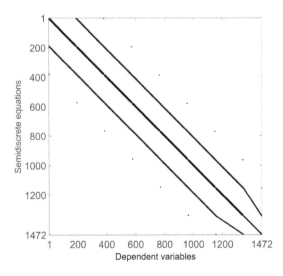

FIGURE 6.42

Frank spheres problem. Pattern of the ODE system matrix with 33,216 non-zero positions resulting from quadratic approximations in each of the 512 finite elements.

The tolerance values (absolute and relative) to the ODE integrator were fixed to 10^{-7}. We solve this problem for $t_0 \leq t \leq t_0 + 0.1$. The penalty constants used are

$$c_1^{m,j} = 10^{-6}, c_2^{m,j} = 0, c_3^{m,j} = c_4^{m,j} = 10^{-3}, c_5^{m,j} = c_6^{m,j} = 10^{-7}$$

To obtain the numerical solution, the integrator performed 375 steps; tried, unsuccessfully, 163 time steps; required 7022 evaluations of the function defining the second member of the ODE system; has defined 123 Jacobians and solved 1096 systems of algebraic linear equations, performing 254 *LU* decompositions. The total computational time spent to complete all calculations was an hour and forty-six minutes.

Figure 6.43 shows the profiles of the solution and the evolution of the spatial mesh at different time instants: $t = t_0 + 0.005$, $t = t_0 + 0.05$, and $t = t_0 + 0.1$. It can be seen that the vertices located on two consecutive circles, that are interior to the spatial domain, start rotating in opposite directions. This movement begins at the interior nodes that are closer to the moving boundary and spreads, in time, to the vertices located at the circles that are progressively more distant from the interface, which may cause the mesh to become progressively less adaptive.

In Figure 6.44, solution cuts may be observed. Slices are taken along $y = 0$, with $x > 0$, at three different time instants, $t = t_0$, $t = t_0 + 0.05$, and $t = t_0 + 0.1$.

Figure 6.45 shows the evolution of liquid-solid interface. The numerical solution obtained by MFEM and drawn with a continuous line, practically coincides with the corresponding curve obtained analytically, represented by a dashed line.

Table 6.3 shows the accuracy of the numerical results obtained depending on the number of finite elements of the spatial grid. The spatial derivatives in the discretized Equation 6.8 were estimated as the arithmetic mean of the local derivatives in finite elements that constitute the support associated to the vertex. It should be noted that these local derivatives may differ by several orders of magnitude. So, in Equation 6.8 is also used the value of $\left. \frac{\partial U}{\partial x_I} \right|_{X_{s_j}}$ given by the derivative of the exact solution. For each instant t such as, $t - t_0 \in \{0.005, 0.01, 0.03, 0.05, 0.07, 0.1\}$, the accuracy was evaluated by calculating the error of the approximate solution in the L^∞ norm. The third and fourth columns of the table show the maximum of these values, considering all the interpolation nodes,

$$E_\infty \left(\boldsymbol{\xi}^k, \cdot \right) = \left| u \left(\boldsymbol{\xi}^k, \cdot \right) - U \left(\boldsymbol{\xi}^k, \cdot \right) \right|_{L^\infty}$$

and only the vertices,

$$E_\infty \left(\boldsymbol{X}_j, \cdot \right) = \left| u \left(\boldsymbol{X}_j, \cdot \right) - U \left(\boldsymbol{X}_j, \cdot \right) \right|_{L^\infty}$$

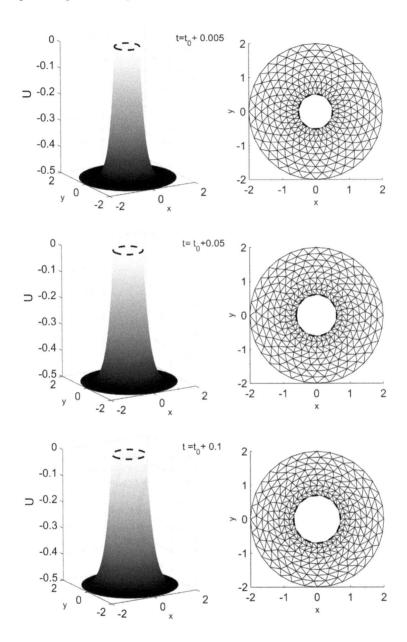

FIGURE 6.43

Frank spheres problem. Solution as a function of spatial variable (x, y) (left) and spatial mesh (right) at different time instants, $t = t_0 + 0.005$, $t = t_0 + 0.05$, and $t = t_0 + 0.1$ with quadratic approximations.

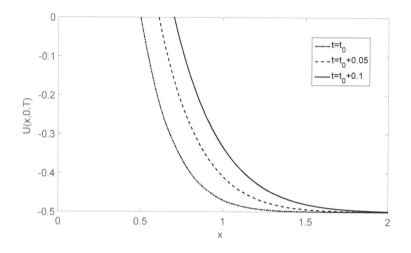

FIGURE 6.44
Frank spheres problem. Slices, along $y = 0$, of MFEM solutions at different time instants.

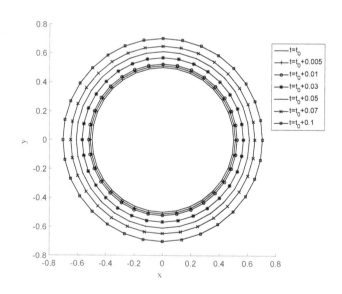

FIGURE 6.45
Frank spheres problem. Interface position, at different times.

TABLE 6.3

Frank spheres problem. Accuracy versus the number of
finite elements.

FEs	$\frac{\partial U}{\partial x_I}$	$E_\infty\left(\boldsymbol{\xi}^k, \cdot\right)$	$E_\infty\left(\boldsymbol{X}_j, \cdot\right)$	$E_\infty(R)$
32	exact	2.03×10^{-2}	1.72×10^{-2}	1.94×10^{-8}
128	exact	1.61×10^{-2}	7.28×10^{-3}	1.17×10^{-8}
288	exact	6.96×10^{-3}	4.24×10^{-3}	2.05×10^{-8}
512	estimated	9.80×10^{-3}	8.19×10^{-3}	4.30×10^{-3}
512	exact	3.00×10^{-3}	2.04×10^{-3}	4.10×10^{-8}
800	estimated	7.02×10^{-3}	6.24×10^{-3}	3.49×10^{-3}
800	exact	2.06×10^{-3}	1.47×10^{-3}	8.68×10^{-8}

The results presented in Table 6.3 indicate that the maximum error decreases with the increase of the number of finite elements. Consequently, the
choice of an initial finite element mesh with *good definition* in the neighborhood of the moving boundary can be crucial in the accuracy of the numerical
solution. In the last column, we display the maximum error in the radius $R(t)$,
calculated on all the vertices belonging to the moving boundary. We can conclude that the error is less than the tolerance required of the integrator, when
$\partial U/\partial x_I$ is obtained from the analytical solution of the problem. Otherwise,
using the estimated spatial derivatives, the error is of the same order of mag-
nitude of the values obtained for the maximum error in global nodes and at
the vertices of the mesh grid.

In this example, we are also interested in finding if the position of the fixed
boundary of the spatial domain used in the simulation affects the movement
of the moving boundary. So, we perform several simulations considering an increasing radius R_F. The maximum error of the approximate solution obtained
through the MFEM improves slightly. For example, using the same initial data
from the third run in Table 6.3 and changing only the fixed boundary radius
to $R_F = 4$, the maximum value of the error is 5.06×10^{-3}. So it can be concluded that, at least for the simulations reported in Table 6.3, the influence
of the position of Γ_F in the movement of liquid–solid interface is negligible.

In Figure 6.46, we present the approximate solution, a cubic piecewise
polynomial, using a mesh of 512 finite elements. The other initial data are
exactly the same as those considered in the first run presented. The computational time was approximately forty-one hours. Passing from quadratic to
cubic approximations, the maximum error does not decrease, so polynomial
approximations of degree two appears to be sufficient to describe the solution
at each instant.

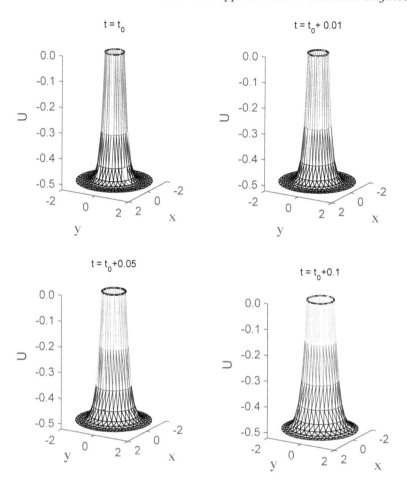

FIGURE 6.46
Frank spheres problem. MFEM solution at different times, $t = t_0$, $t = t_0+0.01$, $t = t_0 + 0.05$, $t = t_0 + 0.1$, and projected mesh, cubic approximations.

6.8   Solidification with Transition Temperature Range

In this section, the MFEM is applied to simulate a fluid $\rightarrow$ mushy $\rightarrow$ solid phase-change system, using the two-dimensional solidification model presented by Gupta [30, 29]. Initially, the spatial domain occupied by the liquid is a rectangular region

$$\Omega = \{(x,y) \; : \; 0 \leq x \leq 1, \; -a \leq y \leq a\}$$

where x is the space coordinate along the length normalized by the length of the rectangle and y is the space coordinate along the breadth of the rectangle normalized by the length of the rectangle. Abruptly, at reduced time $t = 0$, the temperature of the environment adjacent to the boundary $x = 0$ decreases to values below the melting temperature, and so the solid begins to appear.

In this model it is assumed that the transition from liquid to solid phases takes place over a transition dimensionless temperature range $[1, T_{ml}]$, where $T_{ml} > 1$ is the dimensionless temperature of the liquid–mush boundary. If the temperature of the material, at some point in the spatial domain, is less than one then it is in a solid form. On the other end, if the temperature is greater than one but less than T_{ml}, then there is an unstable coexistence of both, liquid and solid phases. If the temperature exceeds T_{ml}, then the material is in a liquid state. Besides, we assume that the system has a plan geometry. The solid-mush interface, which is defined by $x = R_1(y, t)$, separates the solid region from the transition region. The mush-liquid interface, defined by the curve $x = R_2(y, t)$, is the boundary between the transition region and the region occupied by the liquid. An extended formulation of the problem in the three-phase case is given below [30]. The normalized dependent variables are the normalized temperature, T_s, T_m, and T_l, where the subscripts refer to solid, mush, and liquid phase, respectively. In dimensionless form, the mathematical model of this solidification problem for $t > 0$ is defined by

$$\frac{\partial T_s}{\partial t} = \alpha_s \left(\frac{\partial^2 T_s}{\partial x^2} + \frac{\partial^2 T_s}{\partial y^2} \right), \quad 0 < x < R_1(y, t) \quad (6.147)$$

$$\frac{\partial T_m}{\partial t} = \alpha_m \left(\frac{\partial^2 T_m}{\partial x^2} + \frac{\partial^2 T_m}{\partial y^2} + \frac{\lambda}{\beta_1} \frac{\partial f_s}{\partial t} \right), \quad R_1(y, t) < x < R_2(y, t) \quad (6.148)$$

$$\frac{\partial T_l}{\partial t} = \alpha_l \left(\frac{\partial^2 T_l}{\partial x^2} + \frac{\partial^2 T_l}{\partial y^2} \right), \quad R_2(y, t) < x < 1 \quad (6.149)$$

where α is the dimensionless thermal diffusivity. Boundary conditions are given by

$$\left. \frac{\partial T_s}{\partial x} \right|_{x=0} = F_p(y, t), \quad t > 0 \quad (6.150)$$

$$\left. \frac{\partial T_s}{\partial y} \right|_{y=\pm a} = 0, \quad 0 < x < R_1(\pm a, t), \quad t > 0 \quad (6.151)$$

$$T_s = T_m = 1, \quad x = R_1(y, t), \quad t > 0 \quad (6.152)$$

$$\left. \frac{\partial T_m}{\partial y} \right|_{y=\pm a} = 0, \quad R_1(\pm a, t) < x < R_2(\pm a, t), \quad t > 0 \quad (6.153)$$

$$T_m = T_l = T_{ml} \quad x = R_2(y, t), \quad t > 0 \quad (6.154)$$

$$\left. \frac{\partial T_l}{\partial y} \right|_{y=\pm a} = 0, \quad R_2(\pm a, t) < x < 1, \quad t > 0 \quad (6.155)$$

$$\left.\frac{\partial T_l}{\partial x}\right|_{x=1} = 0, \quad t > 0 \tag{6.156}$$

where $F_p(y, t)$ is a dimensionless prescribed flux at $x = 0$, and initial conditions are

$$T_m|_{t=0} = g(x, y) \leq T_{ml}, \quad 0 \leq x \leq R_2(y, 0), \quad g(0, y) = 1 \tag{6.157}$$

$$T_l|_{t=0} = g(x, y) \geq T_{ml}, \quad R_2(y, 0) \leq x \leq 1, \quad g(R_2(y, 0), y) = T_{ml} \tag{6.158}$$

where $g(x, y)$ is the dimensionless initial temperature. The movement of the solid-mush interface at $x = R_1(y, t)$ gives

$$\lambda C_1 \frac{\partial R_1}{\partial t} = \left[1 + \left(\frac{\partial R_1}{\partial y} \right)^2 \right] \left(\frac{\partial T_s}{\partial x} - \beta_1 \frac{\partial T_m}{\partial x} \right) \tag{6.159}$$

and the movement of the mush-liquid interface at $x = R_2(y, t)$ is defined by

$$\lambda C_2 \frac{\partial R_2}{\partial t} = \left[1 + \left(\frac{\partial R_2}{\partial y} \right)^2 \right] \left(\beta_1 \frac{\partial T_m}{\partial x} - \beta_2 \frac{\partial T_l}{\partial x} \right) \tag{6.160}$$

where C_1 is solid fraction at the solid–mush boundary and C_2 is solid fraction at the liquid–mush boundary, β_1 is the reduced thermal conductivity of the mush by the thermal conductivity of the solid, and β_2 is the reduced thermal conductivity of the liquid by the thermal conductivity of the solid and λ a dimensionless parameter.

For the numerical simulation presented, we consider

$$F_p(y, t) = 1 - 0.5y^2 \tag{6.161}$$

$$f_s(x, y, t) = \frac{C_1 \left(T_{ml} - T_m(x, y, t) \right) - C_2 \left(1 - T_m(x, y, t) \right)}{T_{ml} - 1} \tag{6.162}$$

$$g(x, y) = 1 + 0.2x + 0.05xy + 0.1x^2y^2 \tag{6.163}$$

Also, according to [30], we used the following values for the various parameters of the model, $C_1 = 0.9$, $C_2 = 0.1$, $a = 1$, $\alpha_s = 0.6$, $\alpha_m = 0.48$, $\alpha_l = 0.42$, $\beta_1 = 0.8$, $\beta_2 = 0.7$, $\lambda = 0.6$, and $T_{ml} = 1.05$.

Once again we shall define a starting approach to the proposed model which has two phases with null measure. We shall consider that, initially, there exists a very small solid phase, with width equal to $R_1(y, 0) = 1.5 \times 10^{-4}$. Moreover, the movement of the interfaces is from the left to the right of the spatial domain. So, numerical computations were stopped when any point of the mush-liquid interface reaches a certain maximum value, in the neighborhood of $x = 1$, but less than one. The vertices belonging to a moving boundary were imposed to move only through straight lines parallel to the x-axis. The initial condition for temperature in the solid phase was defined constant and equal to one.

The numerical results were obtained in the time interval between $t = 0$ and $t = 1.34$, using quadratic polynomials to approximate the solution in each of the 144 finite elements of the spatial mesh. Spatial domain decomposition is implemented by the introduction of seven moving nodes describing the position of each of the internal moving interfaces. Figure 6.47 shows the meshes on the three regions delimited by the fixed and moving boundaries at time instant $t = 0.01$. In each region, 48 finite elements are used. On the right, we can see a magnification of the space domain showing the position of the two interfaces (dashed lines) at the same time instant.

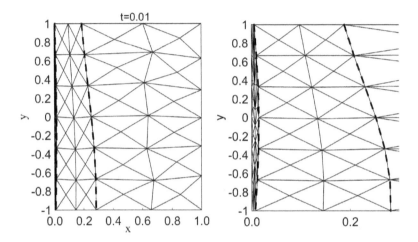

FIGURE 6.47

Solidification with transition temperature range. Space domain discretization at $t = 0.01$ (left). Positions of the moving interfaces (dashed lines) at $t = 0.01$ (right).

The initial discretization is defined by 33 nodes uniformly distributed in the spacial domain occupied by each phase and originates a time-dependent ODE system with 414 equations. The matrix of coefficients of this system has the configuration displayed in Figure 6.48 and it has a degree of sparsity bigger than 95.7% with 7326 non-zero positions.

For the numerical simulation we choose the ODE solver tolerances as 10^{-7}. The values of the penalty constants considered are $c_1^{m,j} = 10^{-6}$, $c_2^{m,j} = 0$, $c_3^{m,j} = c_4^{m,j} = 10^{-3}$, and $c_5^{m,j} = c_6^{m,j} = 10^{-7}$. The statistics relating to the ODE solver are: 4709 successful steps, 2371 steps attempted failed, 45,729 evaluations of the function defining the right side of the ODE system, 797 Jacobian evaluations, 12,045 systems of algebraic linear equations, 3451 LU factorizations. The computation time to complete all the calculations is about two hours and twenty minutes.

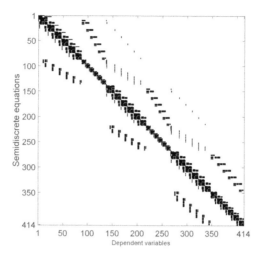

FIGURE 6.48
Solidification with transition temperature range. Nonzero elements pattern of
the ODE system mass matrix.

Figure 6.49 illustrates the evolution of the two interfaces. Our results show
that the new algorithm, based on the MFEM, is able to efficiently describe
the position of the two moving boundaries.

Figure 6.50 plots slices of the approximate solution at $t = 1$, along $y = k$,
with $k \in \{0, 0.3, 0.5, 0.7, 1\}$. The solidification front position is highlighted by
asterisks and circles mark the separation between the liquid and the intermediate phase transition, which is sandwiched between stable solid and stable
liquid regions.

Figure 6.51 shows the temperature histories at three different points, the
two corners of the boundary $(1, 1)$ and $(1, -1)$ and the midpoint of the boundary where the solidification starts, $(0, 0)$. In the two corners located in the
liquid phase, there is a marked decrease in temperature at the initial instants
of the simulation. This variation will be attenuated in time and the temperature tends to the constant T_{ml}. At the origin, the temperature presents an
initial variation and then maintains a nearly linear decrease until the end of
the simulation.

Figure 6.52, on the left, represents the approximate solution, piecewise
polynomial of degree two, obtained at four different time instants, $t = 0.3$,
$t = 0.9$, and $t = 1.34$. The evolution of the corresponding spatial mesh is also
represented. Dashed lines mark the two moving boundaries. Spatial nodes in
the transition region keep their positions in time and have a a highly adaptive
behavior in the solid region, where the solution has a higher variation.

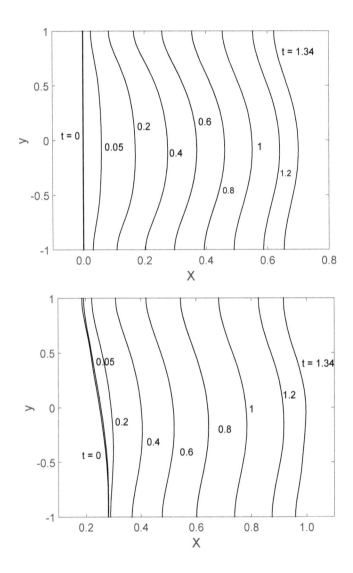

FIGURE 6.49
Solidification with transition temperature range. Solid-mush interface position, R_1, (top) and at mush-liquid interface position, R_2, (bottom) at different instants of time, from $t = 0$ to $t = 1.34$.

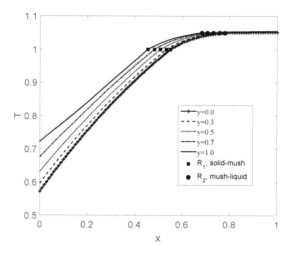

FIGURE 6.50
Solidification with transition temperature range. Dimensionless temperature at $t = 1$ for different values of y as a function of x component of the space variable.

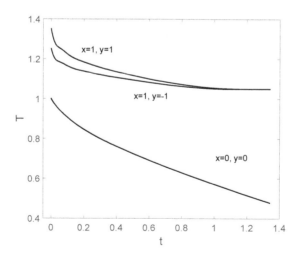

FIGURE 6.51
Solidification with transition temperature range. Dimensionless temperature histories as a function of time variable at three points of the spatial domain, $(0,0)$, $(1,1)$, and $(1,-1)$.

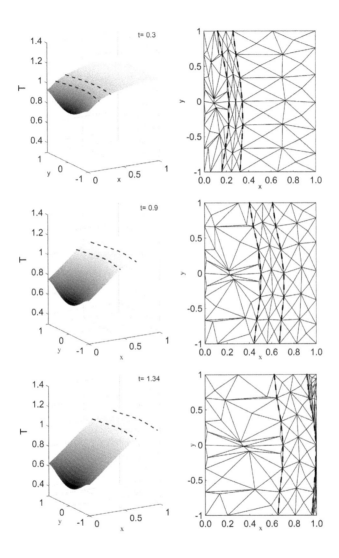

FIGURE 6.52

Solidification with transition temperature range. Dimensionless temperature profiles as a function of space variable in the three phases (left) and the evolution of the finite element mesh (right) for different time instants.

6.9   Conclusions

The MFEM for solving moving boundary problems in one and two spatial domains has been presented. The algorithm is based on the MFEM formulation presented in Chapter 2. The implementation developed takes advantage of the use of various meshes to discretize the spatial domain. Through a spatial domain decomposition, the MFEM can be modified to accommodate the moving boundary. Seven examples in 1D and 2D spatial domains where one or more moving interfaces are present illustrate the derived MFEM for moving boundary problems. The method is found to be efficient and accurate and can also be applied to other problems with moving boundaries arising in modeling chemical engineering systems. However, numerical complications can occur when the spatial domain decomposition leads to a spatial region of null measure. This problem is overcome by choosing an accurate starting solution for the problem. We must define the starting solution in a spatial domain consistent with the MFEM and close enough to the real problem. Our strategy is to obtain the starting solution from numerical approximations of the boundary and initial conditions, or eventually a pseudo-steady-state solution, should be considered to produce an accurate starting solution. The low CPU times indicate a very inexpensive procedure, in particular for 1D moving boundary problems. Numerical results clearly show MFEM provides an accurate and reliable means of solving moving boundary problems.

Bibliography

[1] J. Crank. *Free and Moving Boundary Problems*. Clarendon Press, Oxford, 1984.

[2] J. Struckmeier and A. Unterreiter. A singular-perturbed two-phase Stefan problem. *Applied Mathematics Letters*, 14(2):217–222, 2001.

[3] Svetislav Savovic and James Caldwell. Finite difference solution of one-dimensional Stefan problem with periodic boundary conditions. *International Journal of Heat and Mass Transfer*, 46(15):2911–2916, 2003.

[4] Jirí Nedoma. Numerical solution of a Stefan-like problem in Bingham rheology. *Mathematics and Computers in Simulation*, 61(3-6):271–281, 2003.

[5] J. Lorenzo-Trueba and V.R. Voller. Analytical and numerical solution of a generalized Stefan problem exhibiting two moving boundaries with application to ocean delta formation. *Journal of Mathematical Analysis and Applications*, 366(2):538–549, 2010.

[6] S. L. Mitchell and M. Vynnycky. Finite-difference methods with increased accuracy and correct initialization for one-dimensional Stefan problems. *Applied Mathematics and Computation*, 215(4):1609–1621, 2009.

[7] S. L. Mitchell and M. Vynnycky. An accurate finite-difference method for ablation-type Stefan problems. *Journal of Computational and Applied Mathematics*, 236(17):4181–4192, 2012.

[8] R. M. Furzeland. A comparative study of numerical methods for moving boundary problems. *IMA Journal of Applied Mathematics*, 26(4):411–429, 1980.

[9] E. Javierre, C. Vuik, F.J. Vermolen, and S. van der Zwaag. A comparison of numerical models for one-dimensional Stefan problems. *Journal of Computational and Applied Mathematics*, 192(2):445–459, 2006.

[10] J. Caldwell and Y. Y. Kwan. Numerical methods for one-dimensional Stefan problems. *Communications in Numerical Methods in Engineering*, 20(7), 2004.

[11] Maria do Carmo Coimbra, C. Sereno, and A. E. Rodrigues. Moving finite element method: Applications to science and engineering problems. *Computers and Chemical Engineering*, 28(5):597–603, 2004.

[12] R. J. Robalo, C. Sereno, Maria do Carmo Coimbra, and A. E. Rodrigues. The numerical solution of moving boundary problems using the moving finite element method. *Computer Aided Chemical Engineering*, 20(C):79–84, 2005.

[13] R. Robalo, Maria do Carmo Coimbra, and A. E. Rodrigues. Modelling time-dependent partial equations with moving boundaries by the moving finite element method. In C.A. Mota Soares, J.A.C. Martins, H.C. Rodrigues, J. A. C. Ambrósio, C.A.B. Pina, C.M. Mota Soares, E.B.R. Pereira, and J. Folgado, editors, *III European Conference on Computational Mechanics*, page 47. Springer Netherlands, 2006.

[14] A. C. Hindmarsh. LSODE and LSODIS, two new initial value ordinary differential equation solvers. *ACM-Signum Newslett*, 15:10–11, 1980.

[15] MATLAB®. *Version 8.4.150421 (R2014b)*. The MathWorks Inc., Natick, Massachusetts, 2014.

[16] M. J. Djomehri and J. H. George. Application of the moving finite element method to moving boundary Stefan problems. *Computer Methods in Applied Mechanics and Engineering*, 71(2), 1988.

[17] Pierre Carabin and Dimitrios Berk. Analysis and modelling of the isothermal and nonisothermal shrinking core model with non-linear kinetics. *Chemical Engineering Science*, 47(9-11):2499–2504, 1992.

[18] B. Duarte and A. Portugal. Moving finite elements method applied to the solution of front reaction models: Causticizing reaction. *Computers and Chemical Engineering*, 19:S421–S426, 1995.

[19] R. B. Jarvis and C. C. Pantelides. *DASOLV a differential-algebraic equation solver*. Technical report. Imperial College., 1992.

[20] V. R. Subramanian, H. J. Ploehn, and R. E. White. Shrinking core model for the discharge of a metal hydride electrode. *Journal of Electrochemical Society*, 147(8):2868–2873, 2000.

[21] Y. Q. Lei, C. S. Wang, X. G. Yang, H. G. Pan, J. Wu, and Q. D. Wang. A mathematical model for the cycle life of hydride electrodes. *Journal of Alloys and Compounds*, 231(1-2):611–615, 1995.

[22] S. I. Barry and J. Caunce. Exact and numerical solutions to a Stefan problem with two moving boundaries. *Applied Mathematical Modelling*, 32(1):83–98, 2008.

[23] F. C. Frank. Radially symmetric phase growth controlled by diffusion. *Proceedings of the Royal Society of London, Series A*, 201:586–599, 1950.

[24] R. Almgren. Variational algorithms and pattern formation in dendritic solidification. *Journal of Computational Physics*, 106(2):337–354, 1993.

[25] M. J. Baines, M. E. Hubbard, and P. K. Jimack. A moving mesh finite element algorithm for the adaptive solution of time-dependent partial differential equations with moving boundaries. *Applied Numerical Mathematics*, 54(3-4):450–469, 2005.

[26] H. Chen, C. Min, and Gibou F. A numerical scheme for the Stefan problem on adaptive Cartesian grids with supralinear convergence rate. *Journal of Computational Physics*, 228(16):5803–5818, 2009.

[27] S. Chen, B. Merriman, S. Osher, and P. Smereka. A simple level set method for solving Stefan problems. *Journal of Computational Physics*, 135(1):8–29, 1997.

[28] J. Papac, F. Gibou, and C. Ratsch. Efficient symmetric discretization for the Poisson, heat and Stefan-type problems with Robin boundary conditions. *Journal of Computational Physics*, 229(3):875–889, 2010.

[29] R. J. Robalo. *O Método de Elementos Finitos Móveis para problemas evolutivos com Fronteira Móvel*. PhD thesis, Universidade da Beira Interior, 2014.

[30] S. C. Gupta. A moving grid numerical scheme for multi-dimensional solidification with transition temperature range. *Computer Methods in Applied Mechanics and Engineering*, 189(2):525–544, 2000.

7

Looking Ahead

Modeling and simulation are powerful tools to understand problems present in all areas of Chemical Engineering such as fluid mechanics, heat and mass transfer, flow in porous media, reaction and separation engineering, etc.

A vast majority of dynamic models require numerical methods to solve the resulting differential-algebraic equations (ordinary differential equations ODEs, ordinary differential algebraic equations ODAEs, partial differential equations PDEs or partial differential algebraic equations PDAEs). There are many textbooks of numerical methods describing a vast range of techniques such as finite differences, finite elements, collocation, perturbation, and others [1, 2].

In our group a workhorse was the orthogonal collocation in finite elements (OCFE) [3] and much work on process simulation of adsorption processes and catalytic reactors was done using home-made programs calling routines such as PDECOL [4], GEAR [5] , GEARIB [6] , LSODE and LSODIS [7], COLNEW [8], COLDAE [9], DASSL [10] and the method of lines [11]. A good reference for differential-algebraic initial-value problems is the book by Brenan et al. [12].

One of us (AER) remembers a conversation with the late Professor Colin McGreavy from University of Leeds in Florianopolis in the middle 1990s. He said that in the future the teaching of numerical methods in chemical engineering departments would decrease since friendly software would be available and students could spend more time on process analysis. He was right; since then packages for dynamic simulation have been developed as gPROMS from PSE [13], ADSIM from ASPEN [14], and MATLAB® [15] now has good routines such as *pdepe* [16] to solve models described by PDEs. The disadvantage however is that students lacking the understanding of the numerical methods may misinterpret computer results.

However, modern methods are required in many situations where moving fronts are encountered such as convection dominated systems or transient processes with concentration/heat waves. This is the case in describing the dynamics of catalytic reactors, adsorption processes, coal gasification, chemical flooding in oil fields, etc. [17, 18].

A major contribution to solve problems where steep fronts develop and change position with time was due to Keith Miller [19, 20], who set the basics of the moving finite element method [21]. The idea is to have a mesh which adapts to concentrate the nodes where they are needed in the abrupt transitions. The

MFEM was implemented in our group first to $1D$ problems using polynomial approximations of any degree and later to $2D$ problems. Of particular interest in chemical engineering are packed beds described by $1D + 1d$ models where there are two scales: the axial bed scale and the radial particle scale. All these situations were treated in this book. The question is when to use MFEM. We know by experience that 90% of the dynamic problems can be solved with available commercial software. Sometimes however one has to find another solution and in some cases the MFEM is adequate for those difficult situations. It is like a RollsRoyce ... not to be used everyday.

A major research area in our group is cyclic adsorption/reaction processes. These include processes such as pressure swing adsorption (PSA), simulated moving bed (SMB), temperature swing adsorption (TSA), electric swing adsorption (ESA), continuous ion exchange (CIX), simulated moving bed reactors (SMBR), parametric pumping (PP), expanded bed adsorption (EBA). A common characteristic of these processes is that they reach a cyclic steady state (CSS) after a number of repeating cycles where the variables characterizing the system change with time within a cycle in a similar manner. The incorporation of MFEM in the numerical solution of such cyclic processes is a challenge for the future.

Computational fluid dynamics (CFD) is now a current tool to "visualize" the flow in engineering systems (reactors, separators, porous media, etc.). The basics are well described elsewhere [22, 23, 24]. Of particular relevance is the validation of CFD results as discussed in an excellent book by Roache [25]. Commercial software is available such as FLUENT [26] with user-friendly graphical interfaces. As someone said CFD couples the two well-established disciplines of fluid mechanics and numerical methods. In most cases the numerical technique used is finite volume element method and it is a challenge to extend MFEM to 3D models.

Bibliography

[1] J. R. Rice. *Numerical Methods, Software, and Analysi*. McGraw-Hill, Tokyo, Japan, 1983.

[2] Bruce A. Finlayson. *Nonlinear Analysis in Chemical Engineering*. McGraw-Hill, 1980.

[3] J. Villadsen and M. L. Michelsen. *Solution of Differential Equation Models by Polynomial Approximation*. Prentice-Hall International Series in the Physical and Chemical Engineering Sciences. Englewood Cliffs, New Jersey: Prentice-Hall, Inc., 1978.

[4] N. K. Madsen and R. F. Sincovec. Algorithm 540: Pdecol, general collocation software for partial differential equations [d3]. *ACM Transactions on Mathematical Software*, 5(3):326–351, September 1979.

[5] A. C. Hindmarsh. *GEAR, Ordinary differential equation system solver*. Report UCIC-30001, Rev. 3, Lawrence Livermore Laboratory, Livermore, 1974.

[6] A. C. Hindmarsh. *Preliminary documentation of GEARIB. Solution of implicit systems of ordinary differential equations with banded Jacobian*. Report UCIC-30130, Lawrence Livermore Laboratory, Livermore, 1979.

[7] A. C. Hindmarsh. LSODE and LSODIS, two new initial value ordinary differential equation solvers. *ACM-Signum Newslett*, 15:10–11, 1980.

[8] U. Ascher, J. Christiansen, and R. D. Russell. Collocation software for boundary-value odes. *ACM Transactions on Mathematical Software*, 7(2):209–222, June 1981.

[9] Uri M. Ascher and Raymond J. Spiteri. Collocation software for boundary value differential-algebraic equations. *SIAM Journal on Scientific Computing*, 15:938–952, 1994.

[10] Linda R. Petzold. *A description of DASSL: a differential/algebraic system solver*. Report Sand82-8637, Sandia National Laboratories, Livermore, 1982.

[11] W. E. Schiesser. *DSS/2 (Differential Systems Simulator, version 2), an introduction to the numerical method of lines integration of partial differential equations*. Lehigh University, Bethlehem, 1976.

[12] K. Brenan, S. Campbell, and L. Petzold. *Numerical Solution of Initial-Value Problems in Differential-Algebraic Equations*. Society for Industrial and Applied Mathematics, 1995.

[13] gPROMS Advanced Process Simulation Platform. *Process Systems Enterprise PSE Ltd, UK*. www.pseenterprise.com, Accessed 11 December 2015.

[14] Aspen ADSIM. Aspentech Technology, Inc., MA, www.aspentech.com, Accessed 11 December 2015.

[15] MATLAB®. *The MathWorks, Inc., MA USA*. www.mathworks.com, Accessed 14 December 2015.

[16] Robert D. Skeel and Martin Berzins. A method for the spatial discretization of parabolic equations in one space variable. *SIAM Journal on Scientific and Statistical Computing*, 11(1):1–32, 1990.

[17] Bruce A. Finlayson. *Numerical Methods for Problems with Moving Fronts*. Ravenna Park Publishing, 1992.

[18] M. E. Denn. *Process Modeling*. Longman, New York, 1986.

[19] K. Miller and R. Miller. Moving finite elements I. *SIAM Journal on Numerical Analysis*, 18(6):1019–1032, 1981.

[20] K. Miller. Moving finite elements II. *SIAM Journal on Numerical Analysis*, 18(6):1033–1057, 1981.

[21] M. J. Baines. *Moving Finite Elements*. Clarendon Press, 1994.

[22] Versteeg and W. Malalasekera. *An Introduction to Computational Fluid Dynamics: The Finite Volume Method*. Prentice Hall, 1995.

[23] J. D. Anderson Jr. *Computational Fluid Dynamics*. MacGraw-Hill, Singapore, 1995.

[24] P. J. H. Roache. *Fundamentals of Computational Fluid Dynamics*. Hermosa Pub, Alburquerque, NM, USA, 1998.

[25] P. J. H. Roache. *Verification and Validation in Computational Science and Engineering*. Hermosa Pub, Alburquerque, NM, USA, 1998.

[26] ANSYS FLUENT. *ANSYS, Inc., PA, USA*. www.ansys.com, Accessed 11 December 2015.

Index

Milton Keynes UK
Ingram Content Group UK Ltd.
UKHW031146141024
449569UK00024B/1035